STRATEGIC GUIDANCE

FOR THE NATIONAL SCIENCE FOUNDATION'S SUPPORT OF THE

ATMOSPHERIC SCIENCES

Committee on Strategic Guidance for
NSF's Support of the Atmospheric Sciences

Board on Atmospheric Sciences and Climate

Division on Earth and Life Studies

NATIONAL RESEARCH COUNCIL
OF THE NATIONAL ACADEMIES

THE NATIONAL ACADEMIES PRESS
Washington, D.C.
www.nap.edu

THE NATIONAL ACADEMIES PRESS 500 Fifth Street, NW Washington, DC 20001

NOTICE: The project that is the subject of this report was approved by the Governing Board of the National Research Council, whose members are drawn from the councils of the National Academy of Sciences, the National Academy of Engineering, and the Institute of Medicine. The members of the committee responsible for the report were chosen for their special competences and with regard for appropriate balance.

Support for this project was provided by the National Science Foundation under Contract No. ATM-0405530. Any opinions, findings, and conclusions, or recommendations expressed in this publication are those of the author(s) and do not necessarily reflect the views of the organizations or agencies that provided support for the project.

International Standard Book Number-13 978-0-309-10349-7
International Standard Book Number-10 0-309-10349-5

Additional copies of this report are available from the National Academies Press, 500 Fifth Street, N.W., Lockbox 285, Washington, DC 20055; (800) 624-6242 or (202) 334-3313 (in the Washington metropolitan area); internet, http://www.nap.edu.

Printed in the United States of America

THE NATIONAL ACADEMIES

Advisers to the Nation on Science, Engineering, and Medicine

The **National Academy of Sciences** is a private, nonprofit, self-perpetuating society of distinguished scholars engaged in scientific and engineering research, dedicated to the furtherance of science and technology and to their use for the general welfare. Upon the authority of the charter granted to it by the Congress in 1863, the Academy has a mandate that requires it to advise the federal government on scientific and technical matters. Dr. Ralph J. Cicerone is president of the National Academy of Sciences.

The **National Academy of Engineering** was established in 1964, under the charter of the National Academy of Sciences, as a parallel organization of outstanding engineers. It is autonomous in its administration and in the selection of its members, sharing with the National Academy of Sciences the responsibility for advising the federal government. The National Academy of Engineering also sponsors engineering programs aimed at meeting national needs, encourages education and research, and recognizes the superior achievements of engineers. Dr. Wm. A. Wulf is president of the National Academy of Engineering.

The **Institute of Medicine** was established in 1970 by the National Academy of Sciences to secure the services of eminent members of appropriate professions in the examination of policy matters pertaining to the health of the public. The Institute acts under the responsibility given to the National Academy of Sciences by its congressional charter to be an adviser to the federal government and, upon its own initiative, to identify issues of medical care, research, and education. Dr. Harvey V. Fineberg is president of the Institute of Medicine.

The **National Research Council** was organized by the National Academy of Sciences in 1916 to associate the broad community of science and technology with the Academy's purposes of furthering knowledge and advising the federal government. Functioning in accordance with general policies determined by the Academy, the Council has become the principal operating agency of both the National Academy of Sciences and the National Academy of Engineering in providing services to the government, the public, and the scientific and engineering communities. The Council is administered jointly by both Academies and the Institute of Medicine. Dr. Ralph J. Cicerone and Dr. Wm. A. Wulf are chair and vice chair, respectively, of the National Research Council.

www.national-academies.org

COMMITTEE ON STRATEGIC GUIDANCE FOR
NSF'S SUPPORT OF THE ATMOSPHERIC SCIENCES

JOHN A. ARMSTRONG (*Chair*), IBM Corporation (retired), Amherst, Massachusetts
SUSAN K. AVERY, University of Colorado, Boulder
HOWARD B. BLUESTEIN, University of Oklahoma, Norman
ELBERT W. FRIDAY, University of Oklahoma, Norman
MARVIN A. GELLER, State University of New York, Stony Brook
ELISABETH A. HOLLAND, National Center for Atmospheric Research, Boulder, Colorado
CHARLES E. KOLB, Aerodyne Research, Inc., Billerica, Massachusetts
MARGARET A. LEMONE, National Center for Atmospheric Research, Boulder, Colorado
RAMON E. LOPEZ, Florida Institute of Technology, Melbourne
SUSAN SOLOMON, National Oceanic and Atmospheric Administration, Boulder, Colorado
JOHN M. WALLACE, University of Washington, Seattle
ROBERT A. WELLER, Woods Hole Oceanographic Institution, Massachusetts
STEPHEN E. ZEBIAK, Columbia University, Palisades, New York

NRC Staff

AMANDA STAUDT, Study Director
CURTIS MARSHALL, Program Officer
CLAUDIA MENGELT, Program Officer
ELIZABETH A. GALINIS, Research Associate

Preface

This report comes out at a time of significant opportunities and challenges for the atmospheric sciences. More than ever before, society is recognizing the value of weather, air quality, climate, and space weather forecasts and demanding more sophisticated products and services. The last several decades have brought impressive advances in our knowledge of the atmosphere and the Sun, while illuminating just how much more we have to learn. New observational and computational tools have greatly expanded research capabilities. Yet, the national investment in atmospheric research has remained relatively flat over the past decade, presenting a challenge to those who must decide how best to allocate the available resources.

The National Science Foundation's (NSF's) Division of Atmospheric Sciences (ATM) has asked the National Academies to perform a study that will provide guidance to ATM on its strategy for achieving its goals in the atmospheric sciences. This request reflects a desire by NSF to get a broad view of the health of the atmospheric sciences and to get some guidance on how best to direct resources in the future. In response to NSF's request, the National Academies formed the Committee on Strategic Guidance for NSF's Support of the Atmospheric Sciences. In essence, the committee was asked to consider how ATM can best accomplish its goals of supporting cutting-edge research, education and workforce development, service to society, computational and observational objectives, data management, and other goals of the atmospheric science community into the future. (See Box P-1 for the full statement of task.)

The committee approached its task in two phases. In the first phase, the committee met four times to gather information, interact with the

BOX P-1
Statement of Task for Committee on Strategic Guidance for NSF's Support of the Atmospheric Sciences

At the request of ATM, this committee will perform a study that will provide guidance to ATM on its strategy for achieving its goals in the atmospheric sciences (e.g., cutting-edge research, education and workforce development, service to society, computational and observational objectives, data management). In doing so, the committee will seek to engage the broad atmospheric science community to the fullest extent possible. The committee will provide guidance on the most effective approaches for different goals and on determining the appropriate balance among approaches. In essence, the committee is asked to consider how ATM can best accomplish its mission of supporting the atmospheric sciences into the future. Specifically, this study will consider the following questions:

1. What are the most effective activities (e.g., research, facilities, technology development, education and workforce programs) and modes of support (e.g., individual principal investigators, university-based research centers, large centers) for achieving NSF's range of goals in the atmospheric sciences?

2. Is the balance among the types of activities appropriate and should it be adjusted? Is the balance among modes of support for the atmospheric sciences effective and should it be adjusted?

3. Are there any gaps in the activities supported by ATM and are there new mechanisms that should be considered in planning and facilitating these activities?

4. Are interdisciplinary, foundation-wide, interagency, and international activities effectively implemented and are there new mechanisms that should be considered?

5. How can NSF ensure and encourage the broadest participation and involvement of atmospheric researchers at a variety of institutions?

The study will not make budgetary recommendations. The committee will deliver its results in two parts: (1) a short interim report in fall 2005 that provides a preliminary sense of the committee's overarching conclusions; and (2) a final report by fall 2006 that further considers community input and provides the committee's full analysis and recommendations.

---

broader atmospheric sciences community, and conduct deliberations. At several of these meetings, members of the atmospheric sciences community were invited to share their perspectives on study questions, both in sessions devoted to specific issues and in an "open mike" session when any comments were welcome. In addition, the committee made available a Web site through which members of the community could contribute comments, met with the heads and chairs of the University Corporation for Atmospheric

Research (UCAR) universities, and held town hall sessions at the December 2004 fall meeting of the American Geophysical Union (AGU) and at the January 2005 annual meeting of the American Meteorological Society (AMS). In September 2005, we released an interim report, *Strategic Guidance for the National Science Foundation's Support of the Atmospheric Sciences: An Interim Report*, that provided some preliminary insight in response to the charge from NSF.

The interim report was quite well received by NSF and the broader atmospheric sciences community and served to spark many thoughtful responses. The committee welcomed this feedback received in written form as well as at briefings of the report held for NSF staff, for a fall 2005 meeting of the Board of Atmospheric Sciences and Climate in Boulder, Colorado, and during town hall sessions at the December 2005 fall AGU meeting and at the January–February 2006 annual AMS meeting. In particular, the committee was urged to go further in exploring many of the issues raised in the interim report, such as opportunities for high-risk, potentially transformative research, and to consider some aspects that were not highlighted in that report, including supercomputing and training the next generation of atmospheric scientists. The committee took seriously this input during its deliberations for the second phase of the study. In this final report of the committee, we reiterate many of the findings and recommendations of the interim report, make some modest changes to some of them, and offer several new ones.

Many individuals have assisted the committee in gathering information about the current status and evolution of the atmospheric sciences as well as in organizing meetings. We especially appreciate the efforts of Jarvis Moyers, Jay Fein, and their colleagues at ATM, who graciously accommodated multiple requests for detailed information about the division's activities, budgets, and grants over the past 30 years. Richard Anthes, Susan Friberg, and their colleagues at UCAR and Tim Killeen and his colleagues at the National Center for Atmospheric Research (NCAR) were very helpful in providing information about UCAR/NCAR activities and in planning the committee's meeting in Boulder, Colorado. Most notably, all the input received from the broader atmospheric sciences community has been instrumental in shaping the committee's thinking; we especially acknowledge the comments of the individuals listed in Appendix C.

Finally, it is a pleasure to recognize the outstanding work of the study director, Senior Program Officer Amanda Staudt, who brought to our task both broad knowledge of atmospheric sciences and great skill in the conduct of National Research Council studies. She was ably assisted by Associate Program Officer Claudia Mengelt and Research Associate Elizabeth Galinis.

John Armstrong
Committee Chair

Acknowledgments

This report has been reviewed in draft form by individuals chosen for their diverse perspectives and technical expertise, in accordance with procedures approved by the National Research Council's Report Review Committee. The purpose of this independent review is to provide candid and critical comments that will assist the institution in making its published report as sound as possible and to ensure that the report meets institutional standards for objectivity, evidence, and responsiveness to the study charge. The review comments and draft manuscript remain confidential to protect the integrity of the deliberative process. We wish to thank the following individuals for their review of this report:

David Atlas, NASA Goddard Space Flight Center, Silver Spring, Maryland

Sarbani Basu, Yale University, New Haven, Connecticut

William H. Brune, Pennsylvania State University, University Park

Carol Anne Clayson, Florida State University, Tallahassee

Clara Deser, National Center for Atmospheric Research, Boulder, Colorado

Paul Dusenbery, Space Science Institute, Boulder, Colorado

Delores J. Knipp, U.S. Air Force Academy, Colorado

Venkatachalam Ramaswamy, Princeton University, New Jersey

Gerard Roe, University of Washington, Seattle

Gabor Vali, University of Wyoming, Laramie

Although the reviewers listed above have provided constructive comments and suggestions, they were not asked to endorse the report's conclusions or recommendations, nor did they see the final draft of the report before its release. The review of this report was overseen by George Hornberger, University of Virginia, and Thomas Graedel, Yale University. Appointed by the National Research Council, they were responsible for making certain that an independent examination of this report was carried out in accordance with institutional procedures and that all review comments were carefully considered. Responsibility for the final content of this report rests entirely with the authoring committee and the institution.

Contents

Summary

The reach of atmospheric science extends beyond its foundation in meteorology to encompass a broad range of scholarly pursuits, many with immediate societal relevance. Understanding the atmosphere is fundamental to forecasting severe storms, improving air quality, responding to climate change, and anticipating intense solar storms, among other societal objectives. Today's environmental challenges increasingly require knowledge of how the atmosphere interacts with the oceans, the land surface, the space environment, and with human society. The past 50 years have brought impressive advances in our understanding of atmospheric processes and in our ability to anticipate and prepare for weather and climate events. An ever-expanding suite of observational and computational tools are enabling scientists to look at the atmosphere in entirely new ways. Yet, the opportunities and imperative to advance atmospheric science are more important than ever, especially in the face of changing environmental conditions and even greater societal demand for relevant information and services.

The fact that the Earth's atmosphere is by and large beyond our experimental control fundamentally shapes how atmospheric research is conducted. Improving our knowledge about the atmosphere thus requires a strategy that balances multiple approaches and facilitates the interplay among them. Atmospheric scientists use a mix of direct observations, analysis of these observations, laboratory experiments, numerical modeling, and theory. Ensuring the continued vitality of all of these research methodologies is critical for advancing our understanding of the atmosphere. Likewise, the atmosphere is intimately connected to many other parts of the Earth–Sun system, requiring atmospheric scientists to increasingly seek collaborations across disciplinary boundaries, for example, with solar physicists who examine how solar variability may impact the atmosphere, ecologists who investigate the impact of climate change on terrestrial and marine ecosystems, soil scientists who study gas exchange with the atmosphere, or oceanographers who probe how ocean variability drives the climate system.

The National Science Foundation's (NSF's) Division of Atmospheric Sciences (ATM) supports research to develop new understanding of the Earth's atmosphere and the dynamic Sun. In addition, ATM supports activities to enhance education at all levels, the diversity of the scientific community, and outreach to the public. ATM has asked the National Academies to perform a study that will guide the division's strategy for achieving its goals in the atmospheric sciences (see Appendix A for full statement of task). In response, the Committee on Strategic Guidance for NSF's Support of the Atmospheric Sciences was formed and subsequently authored an interim report released in fall 2005 and this, its final report. The committee reviewed the accomplishments of the atmospheric sciences over the last few decades; it discussed the evolution of the scientific, societal, and institutional context in which atmospheric research is conducted, and it responded to this invitation to offer some guidance on how NSF can best support the atmospheric sciences into the future.

The committee found that ATM is operating in an environment that is ever more cross-disciplinary, interagency, and international, necessitating a more strategic approach to managing their activities in a way that actively engages the atmospheric sciences community. At the same time, ATM must preserve opportunities for basic research, especially projects that are high risk, potentially transformative, or unlikely to be supported by other government agencies. Finally, ATM needs to be proactive in attracting highly talented students to the atmospheric sciences as an investment in the ability to make future breakthroughs. These issues are of importance to ATM broadly, and thus the committee chose to highlight them in this summary.

One important and especially challenging aspect of the committee's charge was to assess the balance among the modes of support employed by ATM. The committee defines balance as the evolving diversity of modes and approaches to ensure the overall health of the enterprise; the use of the word balance does not imply a specific percentage to any particular component. ATM employs a range of modes of support for its activities: grants to individuals and to teams of researchers; small research centers; a large federally funded research and development center, specifically the National Center for Atmospheric Research (NCAR) located in Boulder, Colorado; and the acquisition, maintenance, and operation of observational and computational facilities operated by NCAR, universities, and other entities. The committee finds that the diversity of activities and modes of support are strengths of the program and of our nation's scientific infrastructure. The approach and vision outlined in NAS/NRC (1958) and the "Blue Book" ("UCAR," 1959), which together mapped out the complementary roles of a large national center and the individual investigator university grants program, has served the atmospheric science community well and is the envy of many other scientific communities. The newer modes of support, including multi-investigator awards, cooperative agreements, and centers sited at universities, complement the previously established modes. The present balance is approximately right and reflects the current needs of the community.

RECOMMENDATION: ATM should continue to utilize the current set of modes of support for a diverse portfolio of activities.

The nation is now in a phase of rapid change in graduate education demographics, the role of the United States in the global atmospheric science community, potentially the role of NSF in national atmospheric science funding, and the maturation and interdisciplinary growth of atmospheric science, during what is likely to be a period of constrained budgets. There are now more atmospheric scientists than ever before, doing more diverse and often cross-disciplinary work, at a time when federal opportunities for basic research proposals in the atmospheric sciences are down. Of particular concern is decreasing funding for basic atmospheric research by federal agencies other than NSF, forcing more and more of the community to turn to ATM for basic research funding. This proposal pressure will likely be accompanied by continued demand for investments in observing and computational facilities. Without significant increases in ATM's budget, purchasing these facilities will require trade-offs between investments in "tools" at the expense of funding scientists to conduct research when in truth both will be necessary to advance the atmospheric sciences.

A strategic plan will be essential if ATM is to maintain a balanced, effective portfolio in an evolving programmatic environment. A flexible strategic plan developed by ATM staff with ample community input will enable determination of the appropriate balance of activities and modes of support in the ATM portfolio; help plan for large or long-term investments; facilitate appropriate allocation of resources to interdisciplinary, interagency, and international research efforts; and ensure that the United States will continue to be a leader in atmospheric research. In addition, a strategic planning effort that effectively engages the atmospheric science community will enhance the broad understanding of the rationale behind ATM decisions. The committee understands that the Geosciences Directorate (GEO) is revisiting its strategic plan and urges ATM to coordinate its efforts with those of the directorate. Indeed, the development of a strategic plan for ATM is an excellent opportunity to identify important connections with GEO and with many other parts of NSF, including the Biological Sciences Directorate, the Engineering Directorate, and the Education and Human Resources Directorate. Ideally, the process of developing a strategic plan should be straightforward and revisited at regular intervals. Furthermore, the balance of modes should evolve in the future in a manner that is consistent with strategic planning efforts.

RECOMMENDATION: ATM should engage the atmospheric sciences community in the development of a strategic plan, to be revisited at regular intervals.

Periodic external guidance could help ATM ensure that its activities are continually evolving in a way that meets the needs of the broad atmospheric sciences community. At regular intervals of every five to ten years, an advisory mechanism that engages the broad atmospheric sciences community, with an emphasis on obtaining balanced, objective input, could be quite effective. Some of the issues that such a process could address include the balance and relationships among the range of scientific and societally driven research avenues, among the various modes of support employed by the division, particularly regarding potential inequities in resource distribution between large research centers or facilities and individual investigators, and among the various subdisciplines in atmospheric research.

RECOMMENDATION: ATM should seek strategic guidance from a panel that includes representation from the fields it supports at regular intervals to ensure that its programs are well balanced and continue to meet the needs of the atmospheric sciences community.

With the increasing importance of cross-disciplinary, interagency, and international research to the advancement of the atmospheric sciences, scientists need help to navigate cross-disciplinary, interagency, and international boundaries and overcome the many challenges to successfully finding the support for such work. NSF ATM's public interface, its Web site (*http://www.nsf.gov/div/index.jsp?div=ATM*), provides potential Principal Investigators information on specific, active funding opportunities. However, the ATM Web site does not specifically encourage or guide those who would seek to grow or obtain funding for participation in an interdisciplinary, interagency, or international research program. It lacks any discussion of how to establish a dialog with ATM toward that end and then how links between the ATM and other divisions of NSF, other agencies, or research programs in other countries should be pursued.

RECOMMENDATION: ATM should encourage and guide scientists seeking support to participate in cross-disciplinary, interagency, and international research by developing guidelines and procedures for the process by which individuals and the community initiate a dialog about such research opportunities and then following up with submission of formal proposals.

High-risk, potentially transformative research is instrumental in making major advances in the atmospheric sciences. Thus, it is essential to continually preserve and renew opportunities for this type of research. Among federal science agencies, NSF is a leader in its commitment to support high-risk, potentially transformative basic research. Yet, as modes of support that require larger investments have expanded, and as peer reviewers tend to be risk averse, the opportunities for such funding are perceived as having declined. The atmospheric sciences would benefit if ATM expanded its support of such projects. It is difficult to identify specific steps to address this need, but the situation is sufficiently crucial that ATM should seek new approaches. For example, ATM might consider instituting an explicit solicitation for high-risk research, which would allow these proposals to be judged with more appropriate criteria, make it clear to the research community that the division welcomes such proposals, and ensure that program managers proactively consider supporting high-risk projects. A target of about 10 such grants per year is reasonable, although it is important to realize that opportunities for transformative research may not come every year and sometimes come in spurts. Such an effort might be undertaken as a pilot program and reevaluated after several years to see if it did indeed result in breakthrough concepts frequently enough to be worth continuing.

RECOMMENDATION: ATM should increase the opportunities for targeted grants in support of high-risk, potentially transformative research.

Recruiting and training gifted scientists is perhaps the single most important way to enable the atmospheric sciences to advance more quickly on many research fronts that are important to our nation and the rest of the world. Because relatively few undergraduate programs offer degrees in the atmospheric sciences, talented students may be unaware of career opportunities in the field. Given the societal importance of atmospheric science and the significant national investment in an excellent university infrastructure, a large national center, and other laboratories and institutions, the committee believes that increased efforts to attract more bright students into the field are warranted. In the past, NCAR has offered a fellowship program for graduate students. This effort could be revitalized and expanded as an ATM–universities–NCAR cooperative effort. Such a program could offer graduate student fellows (1) multiyear stipends similar to those for NSF graduate research fellowships and (2) a summer program, conducted jointly by NCAR and the universities near the beginning of the students' graduate studies, to acquaint students with available facilities and research opportunities. A program of this sort, sized to support about 20 new students per year at U.S. universities and advertised widely to undergraduates in related scientific majors (e.g., physics, chemistry, applied math), could be a powerful tool for recruiting top students to the atmospheric sciences.

RECOMMENDATION: ATM should establish a new university–NCAR graduate fellowship program to attract a larger share of the world's brightest students into Ph.D. programs in the atmospheric sciences.

Looking forward, ATM faces the need to marshal a wide range of scientific talents to address the rich intellectual landscape of the atmospheric sciences. The range of the discipline has never been greater and its potential to address many issues of great importance to society has never been more obvious. Chapter 6 of this report includes many additional recommendations for effectively using NSF's resources to advance the atmospheric sciences, from developing new observational tools, making the best use of investments in field programs, and ensuring access to supercomputing resources, to effectively utilizing centers and training the next generation of atmospheric scientists. If ATM continues evolving to meet new challenges, it will be well positioned to advance our understanding of the atmosphere and to apply this knowledge to many issues of societal importance.

1

Introduction

Atmospheric processes have an enormous impact on the lives of Americans and the rest of the world's population. From everyday weather to hurricanes and tornadoes, from the quality of the air we breathe to the integrity of the stratospheric ozone layer, and from the impact of increasing greenhouse gases to that of intense solar storms, understanding the atmosphere is of principal importance. The past 50 years have brought impressive advances in our understanding of all of these processes and in our ability to anticipate and prepare for them. Yet, the imperative to advance atmospheric science is more important than ever, especially in the face of changing environmental conditions and even greater societal demand for relevant information and services.

Research activities in the atmospheric sciences are addressing a wide range of societally relevant topics. For example, more timely tornado warning and more accurate predictions of hurricane frequency, location, and intensity have resulted from research to develop better atmospheric models and observations, and improve our understanding of these phenomena. During the 2005 North Atlantic hurricane season there were an unprecedented 27 named storms, and one of these, Katrina, caused extensive destruction after landfall on the U.S. Gulf Coast. During 2003 and 2004, a record number of tornadoes in the United States caused much loss of property and life. Great advances in severe storm prediction have been made, but we can still do better.

Poor air quality continues to adversely affect the health and life spans of tens of millions of people in the United States and many hundreds of millions worldwide, as epidemiological studies confirm that current urban levels of airborne particulates have serious health impacts. Both forecasting and managing air quality require more precise knowledge of pollutant emissions, transformations, and transport. Furthermore, following the model of numerical weather prediction, new prediction capabilities will be emerging for concentrations of key chemical constituents and for aerosols that impact human health.

Increasing greenhouse gas concentrations are warming the surface of the Earth, with worrisome implications for vulnerable ecosystems, low-lying coastal communities, hydrological systems, the cryosphere, and degraded air quality. Crucial policy decisions involving our energy, industrial, and transportation systems will be made on the basis of increasing capabilities to model the future climate and its response to societal actions. Understanding the atmospheric component of climate variability and change is crucial for making successful projections of future climate conditions.

Intense solar storms impact near-Earth space and the planet's atmosphere, with sometimes dramatic effects on communications and observational satellites as well as ground-based electrical distribution systems. Quantitative models and approaches to forecasting space weather are now reaching the stage similar to the early stages of numerical weather prediction. Our understanding of the Sun now makes it possible to predict future solar cycles on the basis of numerical, physics-based models, and useful predictions that trigger actions to protect satellites, astronauts, and the electrical power grid are emerging.

Farsighted and effective support for the atmospheric sciences will have a crucial impact on needed advances addressing these important problems. During the past 50 years the National Science Foundation's (NSF's) Division of Atmospheric Sciences (ATM) has played a vital role in the advancement of the atmospheric sciences and the enhancement of the field's capabilities to address issues vital to society. Over the next 50 years addressing the pressing atmospheric issues noted above will demand wise and bold investments in the atmospheric sciences. In this report we review the record of ATM activities and the advances they have enabled, assess the current state of NSF-sponsored atmospheric science programs, and discuss actions that we hope will help aid future ATM investments to strengthen our science and enhance its ability to address the atmospherically related problems facing humanity.

ATMOSPHERIC SCIENCES AT THE
NATIONAL SCIENCE FOUNDATION

The fact that Earth's atmosphere is by and large beyond our experimental control fundamentally shapes how atmospheric research is conducted. Atmospheric scientists employ a mix of direct observations of the atmosphere, analysis of these observations, laboratory experiments that seek to re-create atmospheric conditions, numerical modeling, and theory. Atmospheric observations can come from routine weather observations, special field programs of relatively short duration, long-term research observations, and climate observing systems. In many cases our understanding is advanced by continually testing theoretical predictions or simulations of system parameters against observations of these same parameters. By iteratively comparing model results with observations and improving understanding of individual processes, representations of natural physical processes in mathematical models of physical systems, such as the atmosphere, the ocean, or the climate system, are continually improved. An ultimate test occurs when these models are used to predict future behavior of natural systems and are tested against observations. Improving our knowledge about the atmosphere thus requires an approach that balances multiple approaches and facilitates the interplay among them.

NSF is responsible for the overall health of science and engineering across all disciplines and for ensuring the nation's supply of scientists, engineers, and science and engineering educators. The Geosciences Directorate (GEO) of NSF includes the ATM, Division of Earth Sciences, and Division of Ocean Sciences. ATM supports research to develop new understanding of Earth's atmosphere and the dynamic Sun, as illustrated in the organizational chart for the division (Figure 1-1). Over the past six years, ATM has devoted about 30 percent of its budget to supporting the Lower Atmospheric Research Section, 16 percent to the Upper Atmospheric Research Section, 42 percent to the University Corporation for Atmospheric Research and Lower Atmospheric Facilities Oversight Section, and the remaining 12 percent to other activities (including Science and Technology Centers, cross-directorate funding, special activities within GEO, and the division-wide account for midsize infrastructure). ATM's total budget for these activities in 2004 was $238.8 million.

ATM supports activities to enhance education at all levels, the diversity of the scientific community, and outreach to the public. ATM-funded scientists conduct research to address NSF-wide priorities and participate in interagency and international research efforts. ATM employs a range of modes of support for these activities: grants to individuals and to teams of researchers; small research centers (e.g., the Science and Technology Centers); a large federally funded research and development center, specifically the National Center for Atmospheric Research (NCAR) located in Boulder, Colorado; and the acquisition, maintenance, and operation of observational and computational facilities operated by NCAR, universities, and other entities (see also Box 1-1). Approximately two-thirds of the ATM's budget is for science research projects, and the remaining one-third is for facility support (Figure 1-2).

NSF is unique in that its mission explicitly covers federal funding for *basic* research in the atmospheric sciences, which is fundamental to advancing our understanding. Other agencies that fund atmospheric research, such as the National Aeronautics and Space Administration (NASA) and the National Oceanic and Atmospheric Administration (NOAA), have missions that are more applied and mission specific. For example, NASA has funded the *development and application* of technology at least as much as the *use* of the technology in doing basic research; in particular, it has supported the development of satellite sensors, which are used as platforms for probing Earth's weather and climate. NOAA has funded projects with specific missions directed at climate and weather products and services, such as acquiring data to be used in numerical forecast models, to develop and improve weather forecasting models and techniques, and has developed and maintained networks of observing systems in support of them.

NSF's primary role of funding basic research in the atmospheric sciences is very important. Not all basic scientific discoveries are immediately useful to society, but many of the objectives of basic research, when realized, can

FIGURE 1-1 Organizational chart for ATM.

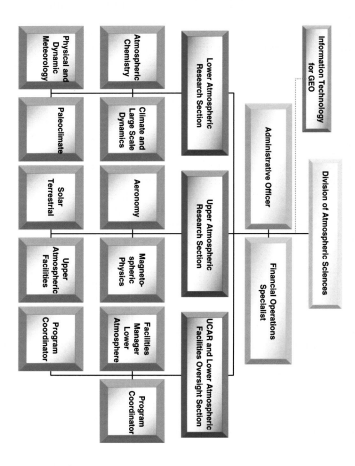

**BOX 1-1
Clarification of Terminology**

The committee is asked to evaluate the "activities" and "modes of support" ATM uses to achieve its goals for supporting the atmospheric sciences. For the purposes of this report, the committee defines these terms as follows:

Goals: The overarching objectives of NSF in supporting the atmospheric sciences, including cutting-edge research, education and workforce development, service to society, computational and observational objectives, and data management.

Activities: The pursuits taken to achieve the goals, including theoretical and laboratory research, field measurement programs, technology development, education and workforce programs, product development, and outreach.

Modes of support: The programmatic tools NSF employs to support the activities, including support for individual or multiple Principal Investigators (PIs), small centers, large national centers, cooperative agreements to support facilities, and interagency programs.

Balance: The evolving diversity of modes and approaches to ensure the overall health of the enterprise. The use of the word "balance" does not imply a specific percentage to any particular component.

Occasionally in this report, "approaches" is used to refer to the collection of activities and modes of support. There are ambiguities in classifying some efforts as activities versus modes of support. For example, field programs are discussed both as an activity that is typically supported by a collection of grants to individual or multiple PIs and as a mode of support because NSF has developed some mechanisms specifically for facilitating field programs.

---

ultimately be applied to society's problems. This was discussed thoroughly in the NSF-commissioned report *Technology in Retrospect and Critical Events in Science* (Illinois Institute of Technology, Research Institute, 1968). In reviewing several case studies of technological and applications developments, the authors of that report note that:

What was most significant, however, was that all applications depended vitally, critically, on a long history of basic research, a substantial part of which was non-mission, uncommitted research.

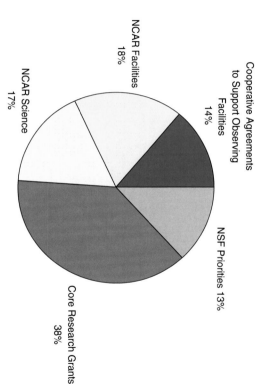

FIGURE 1-2 Expenditure allocations for ATM in fiscal year 2004; total is $238.8 million.

An example from the atmospheric sciences is basic research on severe thunderstorms and tornadoes, which led to a sufficiently improved understanding of them that technology could be used appropriately to better warn the public of impending disaster. Another example is the basic research that provided the foundation for understanding the oceanic and atmospheric conditions associated with El Niño, before the first successful long-lead forecasts during the 1980s. Such advances would not have been possible without the foundation of basic knowledge about the phenomenon's characteristics and behavior. How and why tornadoes or El Niño conditions form must be well known before one can reliably tackle the applied problems of improving warnings or forecasts of them. While applied research can sometimes lead to basic discoveries, in many instances basic research is a prerequisite for successful applied research.

STUDY STRATEGY AND REPORT ROADMAP

To provide NSF's ATM division with the requested guidance the committee solicited broad input from the atmospheric science community in its deliberation in the following ways: (a) Several workshops were organized to invite representatives from the community to provide thoughts on the committee's statement of task; (b) during its initial phase it held several town hall meetings with the purpose of soliciting community input; (c) comments were invited via the committee's Web site; and (d) an interim report was released with the intent to solicit additional community input at subsequent town halls. The committee discussed and considered these comments as part of its deliberative process before drawing its final conclusions.

One important and especially challenging aspect of the committee's charge was to assess the balance among the modes of support employed by ATM. The committee defines balance in Box 1-1 as *the evolving diversity of modes and approaches to ensure the overall health of the enterprise.* In a largely successful program like ATM, the balance is always shifting to reflect changing priorities and opportunities. In this report, the use of the word "balance" does not imply a specific percentage to any particular component. Indeed, there is no way to objectively determine the perfect balance among the modes. That said, the committee took three different tacks to evaluate the balance among the modes and activities supported by ATM in order to identify whether any modifications to the balance were warranted at this time. Chapter 2 examines several major achievements of the field over the past 30 years and to what extent the various modes were important in each. Chapter 3 reviews how the field has evolved over the past 40 years to help us consider whether new modes are needed to address new challenges. Chapter 4 assesses how each mode operates today to identify the strengths and shortcomings of each.

Chapter 5 highlights another major theme of the committee's deliberations: cross-disciplinary, interagency, and international collaborations that are critical for the success of the atmospheric sciences. In the final chapter of the report the committee concludes with its findings and recommendations regarding the overall balance and value of the various funding modes and activities and points to broad areas where attention by NSF is warranted to improve support for the atmospheric sciences.

2

Major Advances in the Atmospheric Sciences

INTRODUCTION

The committee was charged by the National Science Foundation's (NSF's) Atmospheric Science Division (ATM) to assess the balance among the types of activities and modes of support and to make recommendations as to how the balance might need adjustments to ensure the health of the atmospheric sciences into the future. In its interim report the committee recommended that ATM should continue to utilize the current mix of modes of support for a diverse portfolio of activities (i.e., research, observations and facilities, technology development, education, outreach, and applications) (NRC, 2005e). Thus, the committee concluded that the types of activities and the modes of support were appropriate and now addresses the further question of whether the balance among activities or modes should be adjusted.

The committee devoted considerable thought to the appropriate methodology for dealing with this "balance question" in the context of strategic guidance. It decided that the most useful approach would be to create a list of major research accomplishments in the atmospheric sciences, supported at least in part by NSF, and then analyze the role of ATM's modes and activities. The balance would be judged to be in need of adjustment if various modes or activities had in fact *not* been crucial in achieving any of these *major* research accomplishments. Conversely, finding the various modes of support and activities to be well represented among the major achievements of the past decades is good evidence that having a diversity of modes and activities has been a successful strategy. That is, it would show that the mix of modes and activities has contributed to major advances and it would provide evidence that the balance has been adjusted to accommodate new opportunities and needs. Given that the NSF has multiple mechanisms for assuring that the *processes* for granting awards are functioning properly, the committee believes that the present task can be addressed by focusing solely on the major scientific *results* of ATM's programs.

Statistical criteria are often used when judging certain aspects of scientific quality. For example, the number of highly cited papers by field would be a possible approach to identifying the relative effectiveness of fields or modes. This form of measurement is often applied to the contributions of individual Principal Investigators (PIs). However, the committee concluded that such statistical measures are both too imprecise and too beset with complications and biases to be useful for our purposes. They are, moreover, not the type of measurements that are appropriate for other modes of support for the atmospheric sciences.

The highly significant accomplishments selected by the committee are shown in Table 2-1 (in no particular order). It is important to note that this list of major achievements and the selection of case studies was made without prior examination or consideration of the roles the modes played in each of the achievements. While this list is not exhaustive, the committee believes that enlarging the set of major achievements would not change our conclusions regarding the adjustment to the balance between modes and activities. The committee selected case studies from all disciplines within NSF's ATM division. While advances in understanding of climate variability and change are certainly among the most significant accomplishments of the past few decades and a few of the case studies cover aspects of climate

TABLE 2-1 List of Selected Major Achievements in the Atmospheric Sciences

List of Selected Major Achievements

1. Improvements in severe weather forecasting
2. Development of the dropsonde
3. Identifying causes for the Antarctic ozone hole
4. Development of community computational models
5. Development of the wind profiler to observe turbulent scatter
6. Emergence of space weather as a predictive science
7. Understanding the oxidative capacity of the troposphere
8. Identifying the importance of tropospheric aerosols to climate
9. The role of Mauna Loa measurements in understanding the global carbon cycle
10. Improving El Niño predictions
11. Development of helioseismology
12. Reading the paleoclimate record

science, the broad scope of climate science did not lend itself to a case study. The committee refers the reader to the comprehensive assessments of the Intergovernmental Panel on Climate Change (e.g., IPCC, 2001). Some of the case studies focus on advances in tools, while others emphasize break-throughs in knowledge and understanding.

The committee notes also that certain significant achievements in atmospheric sciences lend themselves to quantitative assessment; that is, objective measures of progress over time are available. Quantitative improvement measures are immediately apparent for items 1, 2, 4, and 5 in Table 2-1. Such metrics have been treated extensively in the context of mission-oriented programs, such as the U.S. Climate Change Science Program, in the recent National Research Council (NRC) report *Thinking Strategically: The Appropriate Use of Metrics for the Climate Change Science Program* (NRC, 2005d). However, the committee does not believe it is appropriate or possible to expect all of ATM's major research accomplishments to fit that model of quantitative assessment. In part, quantitative assessment of research is problematic because some of these accomplishments were not planned and therefore do not fit the goal-driven model outlined in this earlier NRC report. Moreover, the type of quantitative measures appropriate for some of the modes is certainly not the appropriate measure for others (e.g., progress in severe weather forecasting vs. productivity and effectiveness of individual PI grants). This raises the issue of how to objectively compare incommensurate measures, a kind of "apples and oranges" problem. This dilemma is another strong reason the committee chose to focus on major accomplishments in addressing the balance issue.

In what follows, summaries are provided of what the committee believes are among the most important research results of the past several decades. These summaries are then analyzed for the ways in which NSF ATM's modes and activities contributed to these achievements, and how, in doing so, they occasionally adjusted the balance between modes and activities. It is clear from our analysis of these case studies that NSF ATM has made effective use of its varied modes of support and that the balance between the modes has evolved over time in response to the needs and opportunities of the field. This chapter also includes many testimonials written by some key participants describing in more personal terms how the achievements were made possible by federal agency or private-sector support. It is important to note that the tenacity and dedication of the investigators, whatever the role of NSF support, was an integral factor in many of the research achievements described. Note that this list is not intended to be exhaustive, but the committee believes it is appropriate to the purpose of this chapter.

CASE STUDIES OF MAJOR ACHIEVEMENTS IN THE ATMOSPHERIC SCIENCES

Case Study 1: Improvements in Severe-Weather Forecasting

It is difficult to trace the one seed that began research conducted by the National Oceanic and Atmospheric Administration's (NOAA's) National Severe Storms Laboratory (NSSL) and the nearby School of Meteorology at the University of Oklahoma (OU), other universities, and the National Center for Atmospheric Research (NCAR), that has led to dramatic improvements in severe-weather forecasting during the last decade or two (Doswell et al., 1993). Using conventional radar and aircraft, NSSL conducted studies of severe convective storms in the 1960s (Bluestein, 1999a). These studies built upon the Thunderstorm Project conducted in the 1940s (Byers and Braham, 1949); further contributions by Chester Newton, Ted Fujita, and Keith Browning in the 1950s and 1960s at the University of Chicago and Air Force Cambridge Research Laboratories (AFCRL) (e.g., Fujita, 1963; Newton, 1963); the Alberta Hail Studies Project; the National Hail Research Experiment; and radar development by Roger Lhermitte, Dave Atlas, Rod Rogers, Alan Bemis, Pauline Austin, and J. Stewart Marshall in France, at AFCRL, Cornell, MIT, and McGill, among others, at the aforementioned institutions and elsewhere. The advent of meteorological Doppler radar in the late 1960s and the development and use of dual-Doppler analysis techniques in the 1970s at NSSL and NCAR provided the most significant leap in the ability to observe the behavior and internal structure of supercells and other convective storms (Davies-Jones et al., 2001).

Equally important as the developments in Doppler radar, the concurrent development of three-dimensional numerical cloud models in the mid 1970s at several universities provided the potential to study the dynamics of severe convective storms by performing controlled numerical experiments. As an example, collaboration between the University of Illinois at Urbana-Champaign and NCAR led to the development of the workhorse "Klemp-Wilhelmson" nonhydrostatic cloud model, which was used for two decades (Wilhelmson and Wicker, 2001). Advances in the capabilities of computers, particularly supercomputers at NCAR, permitted the model to be used for severe-storm research. Pioneering work at NCAR in the early to mid 1980s identified quantitatively the basic environmental parameters supportive of supercells, the most prolific producers of severe weather. At NCAR, expertise was at hand also to physically interpret the mechanisms responsible for supercell formation and behavior (Klemp, 1987). The roles of environmental vertical wind shear and potential thermal buoyancy in producing storm rotation and propagation were elucidated. In the late 1980s and 1990s scientists at NCAR and at universities also investigated the behavior of groups of convective storms, mesoscale convective systems, such as squall lines, again using controlled numerical experiments (e.g., Box 2-1; Rotunno et al., 1988; Weisman and Davis, 1998). The roles of low-level vertical shear and an evaporatively produced cold pool of air in controlling storm structure and evolution were described. The results of these experiments led to an increased awareness and understanding of the conditions leading to damaging, straight-line surface winds.

BOX 2-1
Improving Severe-Weather Forecasting

Morris L. Weisman, Senior Scientist, Mesoscale and Microscale Meteorology
National Center for Atmospheric Research
Ph.D., Meteorology, Pennsylvania State University

I began my scientific career in 1979 when I joined a science group at the National Center for Atmospheric Research (NCAR). My work involved the exploration of convective storms with the goal of improving our ability to forecast severe convective phenomena such as tornadoes and damaging straight-line winds. This work was fostered at NCAR by the unique juxtaposition of talents and resources that Dr. Doug Lilly brought together, ranging from state-of-the-art numerical cloud modeling, developed by Dr. Joseph Klemp in collaboration with Dr. Bob Wilhelmson (University of Illinois), to theoretical expertise, contributed especially by Dr. Richard Rotunno. My research focused on the simulation of convective storms and mesoscale convective systems to reveal the dependence of observed convective structure on preexisting environmental conditions such as thermodynamic instability and vertical wind shear. This fruitful research collaboration has offered new physical insights into a host of significant convective phenomena, including supercells, squall lines, rear-inflow jets, bow echoes, and mesoscale convective vortices. Outside collaborations with university researchers have lead to new insights into, for instance, how supercell storms may interact within a squall line. Other collaborations with National Weather Service forecasters have led to the development of new forecasting techniques such as the improved prediction of convective storm motion.

Many of these advances in knowledge are now used by severe-weather forecasters on a daily basis worldwide. Further, they form the basis for four interactive computer-based learning modules produced by COMET (Cooperative Program for Operational Meteorology, Education, and Training), which are used heavily by the National Weather Service, Air Force, and universities. In all of these endeavors, the synergy of a variety of resources and talents available at an NSF-funded national center such as NCAR has been critical. NCAR combines world-class computational and observational facilities with the theoretical expertise covering the full range of atmospheric phenomena, and provides a high level of access to university and other national and international researchers and forecasters.

Many field experiments have been conducted (Table 2-2), in large part with support from NSF, both for field operations and for development of new instrumentation (Figure 2-1). In the early 1970s, storm-intercept field programs began at NSSL and OU, with funding initially from NOAA (Bluestein, 1999b). Early collaborative annual spring field programs led to a conceptual model of supercells used by spotters and nowcasters, and *in situ* verification of severe weather events that eventually instigated the development of a national network of Doppler radars (NEXRAD) and its implementation in the 1990s. After the radars became operational, the accuracy and lead time of short-term (< 1 h) severe-weather warnings improved greatly.

The object of some of these experiments was to study the details of tornado development and other severe-storm features; the object of others was to further understanding of convective storms in general. In many instances, NOAA provided partial or seed support. Quantitative studies in tornadoes began with the Totable Tornado Observatory (TOTO), built by NOAA, in the early 1980s. Pressure falls associated with mesocyclones and thermal aspects of the rear-flank downdraft were documented. The first portable Doppler radar, developed at the Los Alamos National Laboratory, was used in the late 1980s to estimate the maximum wind speed in tornadoes from Doppler spectra. NSF funded part of these efforts well before instrumented storm-intercept projects were recognized by the community to be scientifically valuable. From these efforts, it was determined that the "thermodynamic speed limit" was usually exceeded, thus pointing to the important role of dynamic pressure gradients near the ground in tornadoes.

A scanning, airborne Doppler radar (ELDORA—ELectra DOppler RAdar) was developed in large part at NCAR and used by university and NCAR scientists to probe supercells during VORTEX (Verification of the ORigin of Tornadoes EXperiment) in the mid 1990s (Bluestein

TABLE 2-2 Some Important Large-scale Field Experiments Conducted in the Last 20 Years

Important Large-scale Field Experiments of the Last 20 Years
Oklahoma–Kansas Preliminary Regional Experiment for STORM-Central (OK-PRESTORM) 1985
Convective Initiation and Downburst Experiment (CINDE) 1987
Cooperative Oklahoma Profiler Studies (COPS) 1989, 1991
Verification of the Origins of Tornadoes Experiment (VORTEX) 1994, 1995
Severe Thunderstorm Electrification and Precipitation Study (STEPS) 2000
International H2O Project (IHOP_2002) 2002
Bow Echo and Mesoscale Convective Experiment (BAMEX) 2003

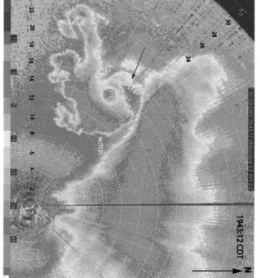

FIGURE 2-1 (bottom) Plan view of a radar image, just above the ground, of the radar reflectivity of a tornadic supercell on May 29, 2004, near Calumet, Oklahoma; from the University of Massachusetts mobile X-band, dual-polarization, Doppler radar. The center of the mesocyclone is located at the hole in reflectivity seen in the left, center. (top) Graduate students from the University of Oklahoma probing a tornado near Hodges, Texas, on May 13, 1989, using a portable, CW (continuous wave)/FM-CW, X-band (3-cm wavelength) Doppler radar from the Los Alamos National Laboratory. Photograph copyright Howard B. Bluestein.

and Wakimoto, 2003). Hitherto unseen details of storm evolution during tornadogenesis were examined for the first time and it was found that surface mesocyclogenesis is not a sufficient condition for tornadogenesis.

At about the same time as VORTEX, several ground-based, mobile Doppler radars were developed for analyzing the structure of the tornado itself. One effort, supported at the University of Massachusetts at Amherst and OU by NSF, led to the development and use of a mobile, high-frequency, ultra-high-resolution W-band radar; the other led to the development of the Doppler-On-Wheels, an X-band radar. The latter was initially supported by NCAR, OU, and NSSL, with some NSF funding, and has been very widely used ever since, not only for severe-storms research, but also in hurricanes and in mid-latitude storms.

Tornadogenesis, which was found to take place on time scales of 10 s or less, in one case appeared to occur when a small-scale bulge in the rear-flank gust front developed, and a small-scale vortex appeared just ahead of it and interacted with a larger-scale low-level mesocyclone. Small-scale shear-induced vortices along the gust front were resolved and hypothesized to potentially play a role in tornado formation. The radial variation of wind speed has been clearly resolved; multiple vortices have been documented, as has the fine structure of weak-echo holes. Since then, other mobile radars have been developed, in part with NSF funding; they promise to add even more significantly to knowledge of tornado structure and formation. It is anticipated that field experiments with these radars, especially during VORTEX-II, which is currently in the planning stage, will further unlock the mysteries of tornado formation and ultimately lead to further improvements in tornado prediction.

The results of the numerical-simulation efforts and the storm-intercept field programs have been applied to severe-storm forecasting through the efforts of COMET, a University Corporation for Atmospheric Research (UCAR) program. A number of forecasters who were supported by university NSF grants as students subsequently became employees at National Weather Service Forecast Offices and/or the Storm Prediction Center.

NSF has not only supported observational and basic theoretical work, which have indirectly led to the advances mentioned above, but it also has funded efforts to improve severe-storm forecasting more directly, through small centers at universities. One of the first of 11 NSF Science and Technology Centers (STCs), the Center for the Analysis and Prediction of Storms (CAPS) at OU, pioneered storm-scale numerical weather prediction in which fine-scale observations, principally from Doppler radar, along with unobserved quantities retrieved from the Doppler-radar observations, are used to initialize cloud-resolving models. CAPS also developed the world's first storm-scale prediction system for massively parallel computers, laying the intellectual and technological foundation for what has become a major area of inquiry including the next-generation Weather Research and Forecast (WRF) model. Other efforts, funded in large part by NSF, continue at NCAR and elsewhere toward perfecting the WRF. An NSF Engineering Research Center, the Center for Collaborative Adaptive Sensing of the Atmosphere, is based at the University of Massachusetts at Amherst, but also has other academic partners including OU, Colorado State University, and the University of Puerto Rico at Mayaguez. It aims to create a distributed, adaptive network of low-power phased-array Doppler radars on existing infrastructure (e.g., cellular towers) to improve severe-weather forecasting and warnings by sensing the region from the ground to 3 km altitude. This effort is jointly funded by the Engineering and Geosciences Directorates at the NSF. Support from and collaboration with industry has also become an important part of these centers. A systems-level testbed of four radars was installed in Oklahoma during January 2006 and will be expanded in the coming years.

This case study illustrates that ATM's diverse portfolio of activities and modes of support were instrumental in the improvements in severe-weather forecasting during the last few decades. In addition to individual PI grants and the support of the large national center, the support of small centers was particularly fruitful for the development of radar technology and numerical modeling tools. In supporting these activities and theoretical work at universities, NSF has also provided essential support for graduate education to many students. Many of them have since become employed not only by NOAA as mentioned earlier, but also as researchers and educators at universities, government laboratories, and at NCAR, thus ensuring the existence of future generations who will further improve severe-weather forecasting.

Case Study 2: Development of the Dropsonde

The remarkable accuracy of hurricane landfall forecasts during the 2005 hurricane season was largely thanks to the use of dropwindsondes (Figure 2-2). Starting with their use in hurricane reconnaissance in the 1960s by the U.S. Navy and Air Force, dropsondes have become an important part of both research and operations, involving NCAR, NSF-supported university research, NOAA, the Air Force, and the private sector. Hurricane reconnaissance using dropsondes dates from the 1960s, when the U.S. Navy and Air Force used Bendix-made dropsondes to sample tropical cyclones in the Atlantic and the Pacific. In 1966, University of Arizona researcher Walter Evans modified a Bendix sonde to sample the electric field in thunderstorms and dropped them from the NCAR Queen Air, introducing dropsondes to the university research community. Then Robert Bushnell and colleagues at NCAR designed a sonde with a downward-pointing pitot tube

FIGURE 2-2 RD-93 aircraft dropsonde.

to measure vertical winds in thunderstorms for the National Hail Research Experiment (NHRE).

The NHRE-inspired design started a decades-long effort of drop-sonde development by a group of NCAR scientists and engineers (in addition to Bushnell, Harold Cole, Stig Rossby, P.K. Govind, Justin Smalley, Dean Lauritsen, Terry Hock, Walt Dabberdt, and Vin Lally), which is also described in Box 2-2. Advances were spurred by the needs of NSF-sponsored field campaigns, international field campaigns, or requests by the Air Force, NOAA, or the Deutsche Luft-und Raumfahrt (DLR); and by improvements in technology. Wind-measuring capability utilizing the

BOX 2-2
Development of Aircraft Dropsondes for
Atmospheric and Hurricane Research

Harold L. Cole, Senior Engineer
National Center for Atmospheric Research
MS, Atmospheric Science, Colorado State University

I was hired by NCAR in September of 1970, on a half-time basis, to support the Viking Meteorology Experiment for the Viking Mars Lander program. Consequently, funding for half of my salary was provided to NCAR by Dr. Seymour Hess of Florida State University, who was the science team leader of the Viking Meteorology Experiment. As the Meteorology Team Engineer, I helped develop the requirements for and test the automatic weather station that led to the first-ever daily weather reports from Mars. During that same time, the other half of my salary was provided by NCAR to work as a Project Engineer for development of the Omega dropwindsonde using the Omega Navigation signals to compute winds. This dropsonde was needed for the upcoming Global Atmospheric Research Program's (GARP's) Atlantic Tropical Experiment (GATE). The NCAR Omega dropsonde system was successfully used during GATE and later in the First GARP Global Experiment (FGGE). The ten aircraft data systems, designed by NCAR and commercially built with NOAA funds for FGGE, were later given by NOAA to the U.S. Air Force and adopted for their hurricane reconnaissance mission.

Subsequently, I served as the Project Manager for the joint U.S.–Canadian development of the Automated Shipboard Aerological Program (ASAP), which was supported by NOAA, NCAR, and the Canadian Atmospheric Environment Service. The ASAP development produced a containerized upper-air sounding system (radiosonde) that can be placed on ships-of-opportunity crossing the North Atlantic and North Pacific oceans allowing radiosonde measurements to be taken over the oceans. The first such sounding system was placed on a Japanese car carrier (M.V. Friendship) in April 1982 and went from Vancouver, British Columbia, to Japan and back. The ASAP program became a WMO-sponsored program in the mid 1980s and continues to this day.

In 1985 the Air Force Hurricane Hunters were starting to have problems with the old Omega dropwindsondes due to rising costs, obsolete parts, and quality control problems. I worked with the Air Force and the Office of the Federal Coordinator for Meteorology to develop a new *smart* (i.e., microprocessor-based), lightweight digital dropsonde that incorporated Loran (Lightweight Loran Digital Dropsonde-L2D2) or Omega (LOD2) windfinding. The Omega version of the dropsonde was adopted by the U.S. Air Force in the early 1990s for its hurricane reconnaissance mission.

In 1987, I developed plans in collaboration with NOAA's Office of Global Programs to put an upper-air sounding system and automatic surface station on Kanton Island in the tropical Pacific due to the TOGA Office's interest in looking at the cause of El Niño and gathering data from the tropical Pacific. Because some of the concepts used on ASAP were directly applicable to the development of a self-contained, easily operated (one person) sounding system, the TOGA Office requested a proposal for the development of the Kanton Island Sounding System (KISS). The system was installed on Kanton Island in August 1988 and continued to operate until after TOGA COARE (~1994). For the follow-up on the TOGA COARE program, which was to understand the interaction of ocean and atmosphere in the warm pool and its role in determining global climate, I led the development of the Integrated Sounding System (ISS). The ISS, which is still one of the major Earth Observing Laboratory (EOL) field project support instruments, combines a Doppler radar wind profiler, an automatic weather station, and a radiosonde sounding system. The data from all three systems are integrated into a data collection, display, and transmission system.

In 1993 the German Aerospace Research Establishment (DLR) contracted with NCAR for a design study to see if it would be feasible to adapt our digital dropwind-sonde system to their new high-altitude research aircraft (STRATO 2C). The study concluded it was feasible if a GPS receiver were used for winds. NOAA had just purchased a new high-altitude research aircraft (G-IV) for their hurricane research with a primary goal to study hurricane development using dropwindsondes. As a result of these two complementary programs and the need for a new NCAR dropwindsonde system, we started the joint development (NCAR/NOAA/DLR) of a new GPS dropwindsonde system called the Automatic Vertical Atmospheric Profiling System (AVAPS). AVAPS was completed and became operational in 1995. The new GPS dropwindsonde system has become the standard for hurricane research by the NOAA Hurricane Research Division and for hurricane reconnaissance by the U.S. Air Force Hurricane Hunters. The new GPS dropwindsonde has made the first wind measurements in the hurricane eye-wall down to the ocean surface and its high-resolution measurements have improved the mean track forecasts by about 30 percent.

The latest contribution to dropsonde technology, developed at NCAR, is the new Miniature In-Situ Sounding Technology (MIST) dropsonde for use with the Driftsonde Balloon system during the Atlantic and Pacific THORPEX program. This new sonde weighs 140 grams instead of 400 grams and is 4.4 cm in diameter and 23 cm long versus 7 cm in diameter and 41 cm long for the aircraft dropsonde. A version of this sonde may someday replace the existing aircraft dropsonde.

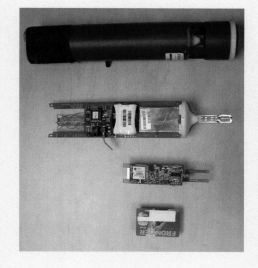

Image: The new MIST sonde and the RD-93 aircraft dropsonde.

Omega navigation system was introduced in time for the Global Atmospheric Research Program (GARP) Atlantic Tropical Experiment (GATE); these sondes were also used in the Global Weather Experiment (GWE) and the MONsoon EXperiment (MONEX) in 1978–1979. This capability was improved through use of the Loran navigation for the Genesis of Atlantic Lows Experiment (GALE 1986) and the Experiment on Rapidly-Intensifying Cyclones over the Atlantic (ERICA) in 1989. Starting in 1994, NCAR partnered with NOAA and DLR to develop the Global Position System (GPS) sonde for deployment by the new NOAA Gulfstream-IV aircraft and DLR's proposed stratospheric research aircraft. While the stratospheric research aircraft was cancelled for cost reasons; DLR has continued to deploy the GPS dropsonde from their Falcon research aircraft. During this same period of time, the response and accuracy of the thermodynamic measurements (temperature, mixing ratio, and pressure) were improved, along with the design of the sonde housing, parachute, and antenna, the conversion to digital mode, and with the major improvements to the onboard data systems. Although the sondes were designed primarily for deployment from aircraft, they were also launched briefly from 80-foot-diameter super-pressure balloons for the GWE (1978–1979).

Manufacturing of the sondes passed between NCAR and the private sector in a stepwise fashion as new versions were designed. The NHRE sondes were manufactured by A.R.F. Products, in Boulder. The Dorsett Electronics Division of LaBarge, Inc. (Tulsa, Oklahoma) manufactured the Omega sondes used in GATE; while Traco, Inc. (Austin, Texas) partnered with NCAR to build the aircraft data system. VIZ (Philadelphia, Pennsylvania) manufactured the thousands of sondes used during the First GARP Global Experiment (FGGE) and MONEX after NCAR tested post-GATE improvements using prototype sondes manufactured by A.R.F. The Loran navigation sondes developed for the GALE and ERICA were manufactured at NCAR. The next-generation sondes, developed for the Air Force and used in the Tropical Oceans Global Atmosphere (TOGA) Coupled Ocean-Atmosphere Research Experiment (COARE) in 1992–1993 and CEntral Pacific EXperiment (CEPEX) in 1993, were manufactured by Radian, Inc. When TOGA COARE PIs from Texas A&M, Colorado State, NCAR, and elsewhere became suspicious of the humidity data from the radiosondes, the dropsondes were useful in verifying the biases. NCAR and Vaisala isolated the cause of the humidity biases. This led to improvements that made the Vaisala radiosondes more robust in tropical environments, providing an enormous benefit to the weather and climate communities. Vaisala also manufactures the GPS sondes used today in meteorological research and hurricane reconnaissance, but NCAR continues to build the data systems.

Improvements in hurricane-track forecasting are largely thanks to the dropsonde (Figure 2-3). In 1982 the National Hurricane Center began to

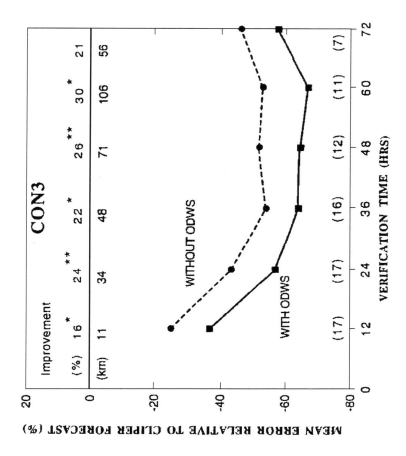

FIGURE 2-3 The average relative errors (in percent) of CON3 with and without the Omega dropwindsondes (ODWs). CON3 is the average of forecasts from three models: HRD's barotropic VICBAR model, NCEP's global spectral model, and the GFDL hurricane model. The numbers just above the zero-skill line are the percentages of improvement of the forecast tracks with ODWs, relative to those without ODWs, where the single- and double-asterisk superscripts indicate significance of this improvement at the 95 and 99 percent significance level, respectively. Numbers just below the zero-skill line are the average track improvements in kilometers, and those in parentheses at the bottom are the numbers of cases for each forecast interval (from Burpee et al., 1996).

assimilate dropsonde data into their operational models on an experimental basis. The results were striking; the reduction in track-error forecasts ranged from 16 to 30 percent—at least as large as the improvement over the previous 20–25 years (Burpee et al., 1996). In 1997, use of dropsondes became operational, with GPS sondes deployed from NOAA's new G-IV aircraft and the Air Force Hurricane Hunter C-130s. The new sonde afforded unprecedented detail in the wind profiles within and around a hurricane.

Such details provide—for the first time—a detailed picture of the near-surface winds in hurricanes, allowing a better estimate of expected damage (e.g., Franklin et al., 2003). James Franklin of the Natural Hurricane Center considers dropsondes the "most important breakthrough" in cutting the uncertainty in hurricane-track forecasts.

Starting with its application to hurricane forecasting and research, the dropsonde has enabled a new mode of observations to support numerical weather prediction—obtaining data where they are most needed. Collecting data for hurricane forecasting has a long history, but in the last decade, the location of more rigorously determined "adaptive observations" are based on ensemble-modeling and adjoint ("backward in time") techniques to identify regions of the atmosphere that could produce the most forecast errors. The Air Force C-130s and the NOAA G-IV fly dropsonde missions over the Pacific to improve forecasts of specific events, using flight plans based on objective, ensemble-based targeting techniques. Such adaptive-observation techniques have been developed and tested in field programs starting with the Fronts and Atlantic Storm Track Experiment in 1997, which involved NCAR, NOAA, some universities, and scientists from France, the United Kingdom, and Ireland.

Improvements in the dropsonde and its use continue. NCAR is developing a new, much lighter GPS dropsonde for deployment from a carrier balloon, to be used for THe Observing system Research and Predictability EXperiment (THORPEX) and the African Monsoon Multiscale Analysis. A modification of this sonde will eventually replace the GPS sonde currently used in research and forecasting. The development of the dropsonde is a clear example of effective partnerships with the private sector. ATM, through the resources provided by a large national center, initiated multiple improvements in the design and effectiveness of the technology, and the private sector leveraged these improvements to manufacture a higher-quality instrument. Coordinated use of the improved technology through many international field campaigns has led to great improvements in understanding and forecast capabilities.

Case Study 3: Identifying Causes for the Antarctic Ozone Hole

In the mid 1980s, a remarkable change in understanding of stratospheric ozone occurred when scientific work by the British Antarctic Survey led by Joseph Farman documented an unprecedented and unexpected depletion of Antarctic ozone (Farman et al., 1985). Ozone appeared to be depleted not by a few percent as models were predicting at that time, but by about a third, and far sooner than any existing theory had anticipated. It was also a surprise that such enhanced depletion was clearly occurring only in the Antarctic, above the world's coldest continent.

Research conducted by scientists worldwide rapidly established industrially produced chlorofluorocarbons as the dominant cause of the remarkable phenomenon that came to be known as the "ozone hole." The enhanced ozone losses in Antarctica compared to other latitudes are linked to the fact that chemical processes that had not been expected can occur on polar stratospheric clouds in that "coldest place on earth."

Policymakers agreed to an international Montreal Protocol to phase out these chemicals, and by the end of the 1990s, global production of these gases had decreased by more than 90 percent. The evolution of scientific understanding of ozone depletion and related policy decisions has since been heralded as one of the most remarkable environmental success stories of the 20th century.

The NSF, including Jarvis Moyers, played many key and unique roles in the scientific support and management that allowed the history of the ozone hole research to progress from observation, to understanding, and to policy in the short space of a few years. NSF was among those responsible for the development of state-of-the-art instrumentation to measure ozone and many key chemicals and dynamical tracers in Antarctica, sponsored by its grants program and its national center. The National Aeronautics and Space Administration (NASA) and NOAA also played important roles in the development of critical measurement capabilities. But those instruments had not been used in the remote Antarctic and the limited available knowledge of the composition, chemistry, and dynamics of the Antarctic atmosphere posed major challenges to establishing the cause of the ozone hole at the time of its discovery.

Within a year after the discovery of the ozone hole, NSF had sent an expedition of four research teams in a National Ozone Expedition to the Antarctic. The 1986 expedition to the Antarctic was strongly led by NSF, with important interagency contributions from NASA, NOAA, and the private sector.

There was a high risk that the work would bear limited if any fruit, but there was also a potential for high payoff. Important strengths included historical approaches to monitoring (e.g., long-term observations of ozone) as well as linkages to instrument development work by NASA, NOAA, and within the NSF astronomy program. Thus, core capabilities in instrumentation, monitoring, and Antarctic logistics were essential to the success of the expedition, which yielded the first measurements showing greatly enhanced chlorine monoxide (de Zafra et al., 1987) and chlorine dioxide (Solomon et al., 1987) in the Antarctic ozone hole. The vertical profile of the ozone depletion was also measured for the first time (Hofmann et al., 1987), providing key evidence for the role of polar stratospheric cloud chemistry. At the time of its discovery, several competing explanations were suggested, including purely dynamical processes, enhanced reactive nitrogen linked to solar activity, and anthropogenic halocarbons (see, e.g., WMO, 1989). The expedition's findings showed that the first two were not consistent with observations, and stand today as among the key initial cornerstones that first established the links between chlorofluorocarbons and the ozone hole.

NSF's Office of Polar Programs (OPP) and ATM worked jointly to make the expedition occur on an unprecedented rapid time scale. Susan Solomon, head project scientist for the National Ozone Expedition, says "NSF's contributions to understanding the ozone hole can only be described as extraordinary. The success can be traced to the dedication of the staff, and their agility in evaluating what needed to be done and why, as well as addressing the enormous operational demands of getting research teams to the Antarctic as quickly as possible."

The following year, NSF-funded investigators also played major roles in another joint interagency campaign, this time using airborne approaches (e.g., Anderson et al., 1989) to further document and demonstrate the key role of anthropogenic chlorine and bromine chemistry on polar stratospheric clouds as the primary cause of the ozone hole. The type of research instruments used in the 1986 expedition are now deployed for monitoring ozone and other chemicals not only in Antarctica but at many sites worldwide, and have contributed to the understanding of Arctic and global ozone depletion as well.

Understanding the Antarctic ozone hole is a case in which the NSF, with significant interagency cooperation, spearheaded an extensive and high-risk research effort to understand and address an issue of vital international importance. Further, the effective partnering of ATM and OPP made it possible to bring the resources and expertise of both divisions to quickly move the research forward.

Case Study 4: Development of Community Computational Models

Development of numerical models for research purposes became widespread starting in the 1960s and 1970s, with individuals from universities, NCAR, NOAA, and elsewhere using numerical codes to simulate the solar interior, synoptic weather, mesoscale weather, severe storms, clouds, and the atmospheric boundary layer. The early models were typically developed by individuals or small groups, with the larger ones run at large computer centers. For the NSF research community, these models were typically run at the NCAR Scientific Computing Facility.

The investment in time and resources to develop a modeling system was so great that community models emerged. The earliest of these, and the most widely used (e.g., Mass and Kuo, 1998), was the mesoscale model first developed at Pennsylvania State University by Richard Anthes and his students, which has evolved today into Mesoscale Model version 5 (MM5; Box 2-3). In the 1980s, Penn State and NCAR jointly developed MM4, and by the late 1980s, NCAR/MMM started supporting MM4 as a true community model, with well-attended community user classes and workshops. The final version, MM5, was released in 1992, improvements continued until 2004, and the last NCAR MM5 tutorial was held in January 2005 (Kuo, 2004). A look at the MM5 parameterization schemes reveals contributions from a broad community—the Blackadar (Penn State University) and Betts-Miller (Betts: CSU and then independent; Miller: ECMWF) boundary layer parameterization schemes, the Grell (University of Miami, University of Washington) and Kain-Fritsch (Penn State University) convection schemes, and the Noah (NCEP, Oregon State, Air Force Weather Agency, NOAA Office of Hydrology) land-surface scheme being some examples. Between 1995 and 2004 the number of users increased from 100 to over 1,100, and the number of institutions using the model increased from 40 to over 560 (NCAR Annual Scientific Reports 1995 and 2004). Other mesoscale models, particularly the Colorado State RAMS model, are widely used, but there are no formal community training workshops.

MM5 is gradually being replaced by the WRF model. Starting in the late 1990s, development of the WRF model began to provide a common modeling system for research and operations and hence to speed technology transfer. The principal large partners are NCAR, NOAA's NCEP and Global Systems Lab, the Air Force Weather Agency, the Naval Research Laboratory, University of Oklahoma, and FAA (http://www.wrf-model.org/index.php), and there has been active participation of the university community (Kuo, 2004). The Beta version of WRF was released in 2000. As in the case of MM5, university PIs have played a part in its development, and workshops and tutorials are held each year (NCAR ASR 2004–2005). The WRF effort now includes two overlapping numerical modeling systems, the Non-hydrostatic Mesoscale Model, operated by NCEP, and the Advanced Research WRF, which is used by the academic community for research and the Air Force for research and operations. While there are significant differences between the two, they still share the same physics packages and software framework.

For mesoscale meteorology researchers, the availability of community models, local access to single- and multiprocessor workstations and gridded analysis, and forecast data enables the university investigator to run mesoscale models at universities for research and education (Mass and Kuo, 1998). The increase in computing power now commonly available is one major factor that made it possible to run such models in a wide array of settings.

Moving from mesoscale models to global models, the Community Climate System Model (CCSM) couples the atmosphere, surface, and

BOX 2-3
Community Models—from Hurricanes to Climate

Richard A. Anthes, President
University Corporation for Atmospheric Research
Ph.D. Meteorology, University of Wisconsin-Madison

In the mid 1960s I was working on the first three-dimensional numerical model of hurricanes. Although I was working for NOAA, NCAR allowed me to use their computer and software to produce one of the earliest visualizations of a three-dimensional hurricane (see graphic). At that time most modelers had his or her own model—usually of specific phenomena such as hurricanes, sea breezes, mountain waves, clouds, or the general circulation (climate). Since the basic equations behind all of these models were the same, I felt that a single model should be able to simulate and forecast disparate phenomena if the resolution, physics, domain size, topographic and surface, and initial conditions were all appropriate.

When I moved to Penn State University in 1971 my students and I began generalizing the hurricane model so that it could be used to study other atmospheric phenomena. Tom Warner, my first Ph.D. student, and I dubbed the emerging model "MM"—for "Mesoscale Model." Over the years, subsequent generations of MM became community models, with the most widely known and used version being the fifth-generation MM, or MM5. Over this time, the MM series was continuously improved by contributions for many universities and laboratories from around the world.

It was always my vision that MM would become a "community model." A community model is one that is used freely and in a cooperative spirit by scientists around the world. A huge amount of work went into developing, supporting, and documenting the basic model, and I thought it would be a waste of time for every student and scientist in the community to develop his or her own model. I also thought that if we all used the same model we would learn more by sharing experiences rather than competing. The NSF provided significant support for the MM effort over the years, as did other federal agencies such as NASA, NOAA, and the Environmental Protection Agency (EPA).

When I went to NCAR in 1981, I brought MM3 with me. Bill Kuo soon joined NCAR and took over the leadership of the MM series. I was Director of the NCAR Atmospheric Analysis and Prediction Division, which had already been developing a Community Climate Model (CCM). The CCM eventually became the Community Climate System Model (CCSM), which has been developed and used by a wide community of climate scientists. The highly successful CCSM has been strongly supported by NSF, with significant support from the Department of Energy (DOE), and this support continues today.

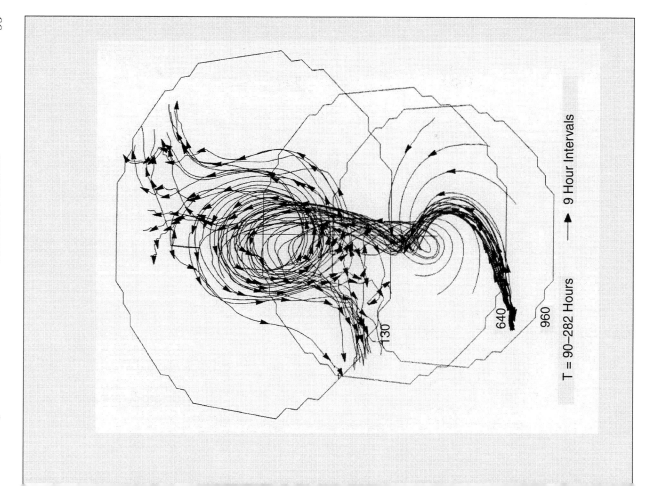

130

640

960

T = 90–282 Hours

⟶ 9 Hour Intervals

oceans on a global scale. Researchers, including many NSF-supported PIs, from 22 universities were participating in the development of the CCSM in 2005, working on land parameterization, atmospheric boundary layer, convection, and radiation schemes, the representation of sea ice, ocean modeling, and some biogeochemistry. Hundreds of scientists meet to discuss their work and plans at the annual CCSM workshop. All model components and the results from major experiments are available on the Web. As of October 2006, there were 297 CCSM publications, authored by individuals at NCAR, universities, and other research entities, frequently in collaboration across these institutions (*http://www.ccsm.ucar.edu/publications/bibliography.html*).

The CCSM is housed at NCAR and has been supported by DOE, NASA, and NOAA as well as NSF. The larger modeling community plays a significant role in its governance (NCAR, 2001; Kiehl, 2004). The need for stable funding, an in-house team of software and hardware engineers, and capability at or near the limits of current computer technology both in terms of speed and storage dictates a centralized operation for these high-end models (NRC, 2001b). Furthermore, dealing with assessments of anthropogenic climate change, ozone, and regional impacts of climate change are best done at a centralized location (NRC, 2001a). Indeed, more than 2PB of data were stored at NCAR from the third IPCC assessment (Kellie, 2004). Projected demands for computing power for coupled climate models outstrip the projected gains from Moore's law (Kellie, 2004). Thus it is likely that CCSM and similar models will be centrally located for the foreseeable future.

Other significant community models are being developed. The Whole Atmosphere Community Climate Model involves NCAR (Atmospheric Chemistry Division, High Altitude Observatory [HAO], CGD) and multiple collaborators from the university community, the private sector, and international partners (Hagan, 2004). An NSF STC, the Center for Integrated Space-Weather Modeling (CISM) is developing a set of coupled codes to characterize the environment extending from the upper atmosphere of the Earth to the surface of the Sun. CISM is based at Boston University, and involves seven other colleges and universities, the private sector, and NCAR/HAO (UCAR Quarterly, 2003).

While there is economy in developing a community model that is improved by contributions from users at multiple universities and government laboratories, it can still become a "black box," that is, used by investigators who do not fully understand the model strengths and limitations. However, for mesoscale models that can be run on university workstations, this can be circumvented by running in a quasi-operational mode and following the successes and failures over a sustained period in a class or for research purposes. Then interesting local effects can be used to understand and then eliminate model shortcomings (Mass and Kuo, 1998). Furthermore, parameterization schemes for such a community model can be tested at a university department, and then shared with the community through inclusion in the new "official" version. Likewise, while it is not realistic to run a coupled climate model at most university departments, it is possible to work on a physical parameterization scheme, an emissions model, a canopy transfer scheme, or another submodel of the community model. Also possible at a university are analysis of model output, comparisons of output to satellite records, and utilization of satellite data as model inputs. Even with these capabilities by individual PIs and their students, frequent workshops and training sessions are needed, and more substantive collaborations involving more substantial PI residence time at NCAR are needed to ensure necessary exchange of information, ideas, and the communication required to foster ongoing collaborations. Such efforts can be used to avoid duplication of effort and ensure more uniform verification procedures (Mass and Kuo, 1998).

The development of community models is another example that illustrates how the balance among the modes of ATM support has fostered a productive relationship among individual university PIs, a large national center that provides capabilities beyond the reach of a university department's resources, and interagency partnerships. In the case of limited domain mesoscale models, technological developments eventually allowed university PIs to conduct research independent of the large national center, yet the center still serves as a maintainer of the "official" version of model code. In the case of coupled climate models, NCAR serves an important role as a provider of computing resources and coordinator of research activities.

Case Study 5: Development of the Wind Profiler to Observe Turbulent Scatter

One of the major successes of funding from the NSF (as well as NOAA) has been the development of the wind profiler. Using radar backscatter of electromagnetic waves in the UHF and VHF from nonthermal fluctuations in the atmospheric refractive index, three-dimensional wind profiles can be obtained nearly continuously and with very high temporal resolution in the troposphere, lower stratosphere, and mesosphere. (Originally these radars were designated as MST radars, referring to the mesosphere, stratosphere, and troposphere). In wind profiling radars, the fluctuations in refractive index arise from clear-air turbulence. Backscattering radars are sensitive to those fluctuations having a scale size of one-half the transmitted wavelength.

The story of the wind profiler begins in the 1940s and 1950s with trying to understand echoes from the clear atmosphere or "angels" observed by radio scientists engaged in radar studies of the lower atmosphere as well as in long-distance over-the-horizon troposphere radio propagation. Many of these original engineers and scientists, through their own curiosity, explored explanations for the observed clear-air echoes. In the late 1950s and 1960s the work of A. W. Friend, David Atlas, Kenneth Hardy, and many others showed that at least some of the echoes were caused by scattering from turbulent irregularities. It was also recognized that specular reflections could also contribute to clear-air echoes at lower frequencies, especially those observed at vertical incidence. The work of Browning (1971) with the Defford radar in the United Kingdom demonstrated the ability to detect lee waves and Kelvin-Helmholtz waves.

In the early 1970s, the focus of research shifted to longer wavelength radars. The pioneering work at VHF was done at the Jicamarca Radio Observatory located in Peru under the direction of Ronald Woodman (Box 2-4). Originally funded by NSF via a special congressional grant, Jicamarca today remains part of a network of NSF-supported high-powered radars that explore the physics of the atmosphere and ionosphere using state-of-the-art radio techniques. The work at Jicamarca culminated in the classic paper of Woodman and Guillen (1974) which theoretically showed and experimentally confirmed the potential of VHF radars to observe the electrically neutral atmosphere. The next step in the development of the wind profiler consisted of radars that were explicitly designed for the purpose of observing the neutral atmosphere.

A flurry of activity ensued in the 1970s and 1980s with funding from NSF and NOAA. Some of the research is summarized in Gage and Balsley (1978), Balsley and Gage (1980), and Atlas (1990). This activity included the design and construction of wind profilers (MST radars) at Sunset (John Green) and Platteville (Ben Balsley), Colorado, and Chatanika, Alaska, as well as further studies at the Jicamarca Radio Observatory and the Arecibo Radio Observatory. The Platteville, Colorado system was the first continuously running, unmanned wind profiler and served as a prototype for the Poker Flat, Alaska wind profiler, which ran continuously from 1979 to 1986. Design and construction of this wind profiler was funded by NSF and served as the major prototype for wind profilers that came thereafter.

Funding for Poker Flat radar was crucial and was a result of the vision of NSF's Ron Taylor, who visited Ben Balsley in Boulder where they discussed preliminary results from the Platteville system. Taylor encouraged Balsley to submit a proposal to NSF, which eventually led to the entire NSF-sponsored wind-profiler program, with the advice ". . . if you don't ask for something big . . . you won't get it." While it was difficult for NSF to justify this project and expense, the ensuing development provided great advancements in operational weather forecasting and the understanding of atmospheric dynamics.

The Japanese quickly became engaged in the profiler development. Following an extended fact-finding visit to Jicamarca they eventually constructed the large VHF radar at Shigaraki, Japan. This powerful and flexible radar with its phased antenna array continues to be adapted to different experimental configurations to study gravity waves, storm development, precipitation, vertical energy coupling, and atmospheric stability. Soon other countries followed with their own wind profiler development, notably throughout Europe, Australia, Taiwan, and India. Major international programs (Middle Atmosphere Program) and workshops (MST workshop series and international radar schools; tropospheric wind profiling conference series) were established and provided opportunities to discuss new scientific understandings gained from the wind profiler measurements.

These scientific developments led eventually to the building of extensive profiler networks for operational use and additional research in the 1980s and 1990s. For example, NSF supported a trans-equatorial Pacific VHF profiler network primarily for troposphere studies associated with El Niño/Southern Oscillation (ENSO). Major information on ENSO, equatorial precipitation, and equatorial dynamics was obtained with this system of five radars (Piura, Peru; Christmas Island, Kiribati; Ponape, ECI; Biak, Indonesia; and Darwin, Australia). The technology began to be transferred from the research community to the operational side of NOAA and the private sector. The NOAA midwest profiler demonstration network was constructed during this period and data are provided routinely to the National Weather Service for use in forecast models as well as for nowcasting. The private sector has provided special-purpose wind profilers for use to monitor low-level winds near airports, and for air quality monitoring systems.

New experimental techniques continue to be developed. The traditional narrow-beam antenna was augmented with interferometry antenna techniques and a radio acoustic sounding system was added to many wind profilers in order to simultaneously measure both winds and temperature. Upper-atmosphere meteor systems were developed as inexpensive add-ons to the wind profiler to obtain winds in the mesosphere and lower thermosphere. Today a wind profiler can not only measure three-dimensional winds but also provide information about wind variability and vertical structure, temperature structure, storm development, divergence and vorticity, momentum and heat flux, turbulence, atmospheric stability, and precipitation.

Of the numerous science problems that have been addressed using wind profiler technology, two highlights are worth mentioning. The first, the Oklahoma–Kansas tornado outbreak on May 3, 1999, is discussed in numerous papers. During this outbreak, the wind profilers were critical in identifying the evolving atmospheric wind patterns, leading to a quick upgrade in the forecast for severe weather. It is estimated that the death toll,

BOX 2-4
The Upper Atmospheric Facility at Jicamarca

Ronald Woodman, Presidente Ejecutivo
Instituto Geofísico del Perú
Ph.D., Harvard

I have enjoyed the sponsorship of NSF for my over 40 years of scientific research. It started in 1966 when I was doing research at Harvard University's Engineering Science Laboratory for my Ph.D. Since then I have spent my career doing theoretical and experimental research related to the use of radars for the remote sensing of the lower and upper atmosphere, from a few kilometers to a few thousand kilometers of altitude. This includes the neutral as well as the ionized atmosphere.

My relationship with NSF has been different than for most researchers. Although about 120 of my close to 130 publications in refereed journals include a well-justified acknowledgement to NSF, in none of them have I been the PI for the corresponding NSF grant, even in the papers where I was the lead author. The reason is that, with a few exceptions, I have been affiliated all of these years with two major radars, the Jicamarca and the Arecibo radars, that are now part of the incoherent scatter radar chain of the Upper Atmospheric Facilities of the Upper Atmospheric Research Section. Both receive full support from NSF through two funding modes, a large core grant to support the general operations and individual grants to the users of the facilities to cover the incremental costs of their particular research. Having been part of the resident staff of both radars has permitted me to make use of the facilities without having to write a proposal to obtain additional funding. I have had the additional advantage during certain periods of having been the Director of the Jicamarca Radio Observatory (1969–1974, 1985–2000) and the Head of the Atmospheric Sciences group at the Arecibo Observatory (1979–1981).

Have I abused this freedom? I don't think so. One reason for this conclusion is contained in the citation for the Appleton Prize of the Royal Society of London that I was awarded in 1999. This citation states that the prize was awarded *"for major contributions and leadership in the radar studies of the ionospheric and neutral atmosphere,"* i.e., for the work I did at both facilities. My peculiar situation illustrates the benefits of NSF's policy with regards to the local scientists at the National Facilities they support. The local scientific staff is envisioned to be the radar experts that help external users with expert advice in the use of the instruments. Additionally, the local scientific staff is envisioned to be responsible for the constant development of the facilities. It is particularly this second function that has allowed me to make the most significant contributions to the field. Of all the work I have done at Jicamarca and Arecibo, it is that related to the development of new capabilities, which makes me feel most proud of my professional achievements. It is for this work that I have received the highest recognition from my peers, including the

Appleton Prize. The capabilities of these two observatories, originally designed and built to measure a few state parameters in the ionosphere are now capable—thanks mainly to the "core" staff—of making important contributions to our observational capabilities and understanding of both the upper and lower atmosphere, capabilities which were not even dreamed by their original promoters.

Ron Woodman, in front of the Jicamarca antenna at the Jicamarca Radio Observatory, which is a facility of the Instituto Geofísico del Perú and is operated under a Cooperative Agreement with Cornell University supported by NSF.

which was 46, may have been as high as 700 within this region of approximately one million people if warnings had not been issued (NRC, 2002b). The second highlight is the original research by Vincent and Reid (1983), who designed a novel experimental setup to observe mesosphere gravity wave momentum flux by using the Buckland Park wind profiler outside of Adelaide, Australia. This research provided the first measurements of mesosphere gravity wave momentum flux and spurred the development of better gravity wave parameterization techniques that are critical for global and specialized models.

The wind profiler story is one of initial radio technique serendipity combined with persistent engineering and scientific pursuit. Initially supported by NSF funding from a visionary program manager, the beginnings of atmosphere radar observations led to a greatly improved understanding of the dynamic atmosphere. Eventually through technology transfer to the private sector, the technique became an important tool in operational forecasting and other applications.

Case Study 6: Emergence of Space Weather as a Predictive Science

Terrestrial meteorology has achieved huge advances over the past 50 years as predictions have become more and more reliable. This predictive ability is certainly one of the great success stories of contemporary science, and it has made a tremendous impact on the lives of everyone. Less well known, and still just in its beginnings, is the emergence of space weather as a predictive science. This success story would not have been possible without the support, encouragement, and vision of ATM, including Rich Behnke, as well as Tom Tascione of the Air Force.

Space weather refers to changes in the space environment that can have an impact on humans and their technology. Storms in space can produce radiation levels that are hazardous to spacecraft and astronauts. Storms in space also produce ionospheric disturbances that can degrade GPS accuracy and interfere with radio communications. Additionally, large conductors (pipelines, power grids) are vulnerable to geomagnetically induced currents that are produced by such storms. While many of the vulnerabilities are new, space weather effects have been around since the mid-19th century when it was noticed that magnetic storms were associated with degraded telegraph operations (e.g., Carlowicz and Lopez, 2002). Today, as dependency on space-based systems increased, power grids become more intertwined, and human exploration of the solar system is considered, the ability to predict space weather has never been more important.

The historical experience with the rise of terrestrial meteorology provides a roadmap for the emergence of space weather prediction (Siscoe, 2006). The Upper Atmosphere section of ATM, being aware of this history, has shepherded this process in the solar and space physics community. Scientists in ATM, working with community leaders, recognized in the early 1990s that space weather prediction was a logical, and needed, product of the basic research that had been funded for decades by NSF, NASA, and other agencies. ATM provided critical leadership for the creation of an interagency plan, the National Space Weather Program (NSWP, 1995), which involved NSF, NASA, NOAA, and the Department of Defense as Co-Chairs, with participation by the DOE, the Department of the Interior, and the Department of Transportation.

The effect of the National Space Weather Program has been dramatic. Since its inception, major documents guiding the research community, such as *Space Weather: A Research Perspective* (NRC, 1997) and *The Sun to the Earth—and Beyond: A Decadal Research Strategy in Solar and Space Physics* (NRC, 2002c), have highlighted the need for improved space weather predictive capabilities. Within NOAA, the Space Environment Center (SEC) was transferred into the NCEP, making SEC part of the National Weather Service. And the space physics community has made space weather a major part of its effort. This development is due in large part to targeted funding within ATM and the success of ATM in participating in crosscutting NSF activities and collaborative programs with other agencies, such as NASA, and with the private sector (Box 2-5).

Within ATM, special solicitations for space weather applications have opened up opportunities for researchers to work on more applied science. Such support is crucial for producing products that can cross the "valley of death" in moving from research to application (NRC, 2000). ATM has also been successful in participating in NSF-wide programs to fund multi-investigator space weather efforts. Two small centers have been established through such means. One is an STC, the CISM, headquartered at Boston University. The other is the Center for Space Environmental Modeling (CSEM) at the University of Michigan, which received support from a Knowledge and Distributed Intelligence grant from the NSF and additional support from the Department of Defense Multidisciplinary Research Program of the University Research Initiative. Both of these centers are creating end-to-end models of the space environment from the Sun to the Earth (e.g., Hughes and Hudson, 2004). While many features of the codes differ, both centers are developing things such as magnetosphere models to be used for near-Earth space weather prediction (see Figure 2-4).

Outside of NSF, ATM has also partnered with NASA's Living with a Star program to solicit proposals to create specific products for space weather prediction, and provided support for the Community Coordinated Modeling Center that makes space weather simulations available to the community at large and which also provides support for validation and metric-based evaluation of the codes. Research by a team led by NCAR

BOX 2-5
Space Weather as a Predictive Science

Louis J. Lanzerotti, Distinguished Research Professor
New Jersey Institute of Technology
Ph.D., Physics, Harvard

I joined AT&T Bell Laboratories after graduate school in the Fall of 1965 to work in both science and engineering related to space. This was in the early days of the space age, and communications satellites were coming to the fore with the first active low-Earth-orbit satellite, Telstar1, having been launched in July 1962. My first job responsibilities were in what is now called space weather (a term that was unknown at that time) and were supported by the company, as was most of my engineering research for nearly four decades. In particular, I was involved with data analysis from the charged particle detectors on Telstar and the construction of a charged particle experiment package for the first geosynchronous satellite ATS-1, which was launched by NASA in December 1966. My first encounters with the NSF were in the very early 1970s when I made a logistics-only proposal to what is now the NSF OPP to install instruments in the Antarctic to conduct both space weather and science of space measurements. These initial Antarctic measurements, geomagnetically conjugate to measurements that we were making in the northern hemisphere in New Hampshire and Quebec, were conducted at Siple Station in collaboration with groups at the University of Maryland and Stanford University. I have been involved with OPP almost continuously ever since with various levels of logistical support until two or three years ago when I joined the New Jersey Institute of Technology. To my knowledge, I have never had direct NSF funding from the ATM to support my efforts on space weather, although I have had logistics support to use, for example, the NSF-supported Sondrestrom radar facility on more than one occasion. It was my industrial support that really mattered for the base of my space weather activities. Nevertheless, the logistics support provided by NSF was critical to allow me to conduct many research investigations that considerably enhanced what I would have otherwise not been able to accomplish. In my view, the mutual leveraging of the industrial and government support over the years was very beneficial for both our company and for our country.

scientists (Dilpati et al., 2004) that led to a major advance in understanding and predicting the solar cycle was supported by joint NSF/NASA funding. There is also a close working relationship with the SEC, and participation in SEC's Space Weather Week, which brings together researchers, forecasters, and customers of space weather predictions. ATM has also had an innovative partnership with the American Geophysical Union, providing seed money for the creation of a new journal, *Space Weather*. This journal provides a publication venue where research, applications, and policy can

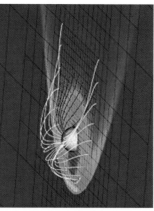

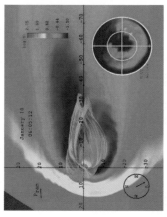

FIGURE 2-4 Visualizations of magnetosphere simulations from CISM (left) and CSEM (right).

mingle in a way that provides a unique community forum (Lanzerotti, 2003).

The emergence of space weather prediction as a field of applied science is an accomplishment in which ATM played the leading role. ATM was in the lead in identifying the fact that progress in basic research had reached the point that one could think about predictive models, some of which are now being transitioned to operations by the SEC. ATM was in the lead in developing an interagency strategic plan, the National Space Weather Program, which has had a significant impact on the agencies and the community. ATM provided the funding space, either on its own or in collaboration with NSF-wide initiatives or other agencies, within which PIs and small centers could advance the state of the art in space weather modeling. And through innovative grants such as the support for *Space Weather*, ATM has provided a safe harbor for an emerging community. The emergence of space weather as a predictive science is a clear success of farsighted ATM leadership in the public interest.

Case Study 7: Understanding the Oxidative Capacity of the Troposphere

The Earth's lower atmosphere has an amazing capacity to oxidize a wide range of chemical compounds emitted by both human-induced and natural processes. Oxidation of most species in the lower atmosphere is driven by reactions of the hydroxyl radical, OH (IGBP, 2003; Prinn, 2003), although some oxidation is accomplished by other radicals, including Cl, ClO, BrO, IO, NO_3, and HO_2, as well as nonradical oxidants such as O_3 and H_2O_2. Oxidation tends to make most airborne chemical compounds more soluble in water, greatly increasing the efficiency of their removal from the lower atmosphere both by precipitation (wet deposition) and contact with surface water, soil, and vegetation (dry deposition). Without this vigorous "oxidative capacity," the lower atmosphere would quickly become choked with anthropogenic and natural pollutants, greatly reducing visibility, degrading the photosynthetic efficiency of plants, and impacting respiration processes in animals and humans. The capacity of the atmosphere to oxidize and efficiently remove the chemical pollution emitted into it is now recognized as one of the planet's major ecological services. Answering the question "Is the 'Cleansing Efficiency' of the atmosphere changing?" has been identified by the International Global Atmospheric Chemistry Project (IGAC) of the International Geosphere-Biosphere Programme (IGBP) as a major challenge for the world's atmospheric chemistry community (IGBP, 2003).

Prior to the early 1970s, the robust oxidation chemistry of the lower atmosphere was not recognized. The fact that the daytime lower atmosphere can be viewed as a photolysis-driven low-temperature flame was simply overlooked. We now have a much better understanding of the complexities of oxidation processes. Oxidizing radicals are now known to be active in gas phase, heterogeneous (gas/surface), and condensed phase reactions, the latter two involving atmospheric aerosol particles and cloud/fog droplets. Further, we now know that, in polluted urban and industrial areas characterized by abundant volatile organic hydrocarbon and NO_x emissions, this chemistry can run rampant, producing the unhealthy levels of ozone and other oxidants, as well as abundant secondary aerosol particles that together characterize photochemical smog (NRC, 1991, 1998b; NARSTO, 2000; IGBP, 2003). The understanding of oxidation chemistry in the lower atmosphere and how this chemistry changes under varying environmental conditions forms the foundation for photochemical models of the lower atmosphere that are critical to air quality assessment.

The photochemical model predictions of oxidative cleansing must be verified experimentally. Since the OH radical is the major cleansing agent, considerable effort has been expended to measure its atmospheric abundance in order to calibrate and test photochemical models. Over the past ~25 years several successful methods to directly measure local concentrations of atmospheric OH as well as the closely coupled radical HO_2 have been developed. Tropospheric measurements of ambient OH have been performed using open-path differential optical absorption spectroscopy (Mount, 1992), chemical ionization mass spectrometry (CIMS; Eisele and Tanner, 1991), and laser-induced fluorescence (LIF; Davis et al., 1979; Wang et al., 1981; Hard et al., 1984; Stevens et al., 1994). LIF measurements that employ atmospheric sampling through a supersonic expansion can also measure ambient HO_2 by using NO to titrate that radical to form OH (Hard et al., 1984; Stevens et al., 1994). These direct OH measurement techniques are now routinely deployed on aircraft and ships as well as at ground sites for major photochemistry-oriented field measurements (Box 2-6). The high time resolution data they provide are utilized, along with data on nitrogen oxide and volatile organic compound concentrations, solar radiation, and other chemical and environmental parameters, to directly test photochemical models. In addition, OH reactivity (the inverse of the local OH chemical lifetime) can now be measured using LIF techniques to trace the decay of an induced OH spike in an ambient air sample (Kovacs and Brune, 2001). Such measurements directly characterize the local atmospheric pollutant loading susceptible to oxidation by OH (Figure 2-5).

The NSF has played a vigorous role in promoting and funding U.S. research on tropospheric oxidative capacity. For instance, in the mid 1980s, the agency took a lead role in assembling a steering committee and selecting task groups to define the elements of the nation's Global Tropospheric Chemistry Program. A key component of that program identified and defined long-term gas phase photochemistry research goals and strategies needed to quantify tropospheric oxidative processes (UCAR, 1986). Since then the agency has consistently sponsored the instrument development, process-oriented field measurements, diagnostic model development and utilization, and basic laboratory experimental and theoretical chemistry and spectroscopy research initiatives outlined in that document. Starting even earlier, in the late 1970s, the agency took a strong lead in supporting direct ambient OH and HO_2 radical measurement techniques, eventually funding or co-funding pioneering LIF instrument development at Georgia Tech, Ford Motor Co., Portland State, and Penn State, as well as supporting the seminal work on CIMS detection at Georgia Tech and NCAR. Currently, field studies and related modeling efforts funded or co-funded by NSF are characterizing the lower atmosphere's oxidative capacity for a full range of ambient conditions from the remote arctic to the world's megacities.

Case Study 8: Identifying the Importance of Tropospheric Aerosols to Climate

The brownish haze associated with many industrial regions and with rural areas subjected to heavy biomass burning is a well-recognized result of human activities. This haze can be transported over long distances to form a regional-scale aerosol layer, such as has been observed in the Arctic, across India and southern Asia, extending from east Asia across the Pacific, and in biomass burning and dust plumes from North Africa that spread over most of the subtropical Atlantic (Ramanathan et al., 2001b). In the late 1990s significant advances were made in understanding such atmospheric aerosol layers and how they affect climate (Box 2-7).

Understanding the sources and fate of tropospheric aerosols has been

BOX 2-6
Checking the Troposphere's Oxidative Capacity

William H. Brune, Professor
Pennsylvania State University
Ph.D., Physics, Johns Hopkins University

It's pretty scary writing your first major proposal in your first faculty position, even if you've been a research associate for ten years. Fortunately, I had reviewed some proposals for the NSF's ATM and thus had learned what makes a good proposal. I decided to focus on measuring the elusive but important tropospheric hydroxyl radical, OH, which I once called the "Howard Hughes of atmospheric chemistry." Our first proposal to develop a new instrument, based on a clever laser technique developed by Portland State University scientists, received good reviews but was too expensive for NSF alone. NSF and NASA split the cost because NASA was also interested in airborne OH measurements. We made rapid progress, only to stumble not once, but twice, during our multi-investigator field campaigns during the early 1990s. Despite less-than-stellar reviews on the renewal proposal, ATM program directors at NSF had the faith to support us.

The reward for this support is ten successful ground-based and seven successful airborne field campaigns in the last decade. The list of measured variables began with the hydroxyl (OH) and hydroperoxyl (HO_2) radicals, but has now expanded to OH reactivity (the inverse of the OH lifetime), naphthalene (a fluke of spectroscopy), and, most recently, HO_2 vertical flux. Often models and measurements agree to within an acceptable uncertainty level, but important systematic discrepancies remain. We are most often asked about our measurement of larger-than-expected nighttime OH, which so far defies our best efforts to find an instrument artifact that might explain this observation. Other discrepancies have more significance for understanding oxidation chemistry. These are the less-than-expected OH in the midday midlatitude middle troposphere, implying slower atmospheric oxidation there, and the less-than-expected HO_2 decrease with increasing nitric oxide (NO), implying greater-than-expected ozone formation at high NO. Some other groups agree with these observations; some do not. Causes for these potentially important discrepancies are being explored with NSF support and other research initiatives.

Our contributions to atmospheric oxidation chemistry, such as they are, would not have been possible without the knowledge, faith, and support of ATM program directors. It is their job to set programmatic priorities within budget constraints and to support investigators' ideas based on reviews from the community and on their own judgment.

a major challenge due to the complexity of their sources, composition, chemical interactions, and physical processing in the atmosphere. Fossil fuel combustion and biomass burning emit particles (e.g., fly ash, dust, and black carbon) and aerosol precursor gases (e.g., SO_2, NO_x, and volatile organic compounds), which form secondary aerosols through gas-to-particle conversion (Ramanathan et al., 2001a). Atmospheric aerosols exist

I have always felt that, although we have stumbled on our own, they have been partners in our successes. They have given us the opportunities to advance the atmospheric sciences, which is what NSF's ATM should be all about.

William Brune atop the ~30 m tower at the Program for Research on Oxidants: Photochemistry, Emissions and Transport site at the University of Michigan Biological Station near Pellston, Michigan, in summer 1998.

in a variety of hybrid structures: liquid droplets, externally mixed (a mixture of particles that each have single chemical compositions), internally mixed (each particle includes multiple chemical components), coated particles, or a combination of all of the above. They are subject to heterogeneous chemical reactions, phase changes, atmospheric transport, and removal from the atmosphere through precipitation or dry deposition to the surface.

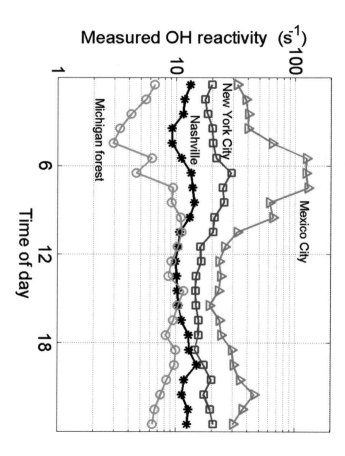

FIGURE 2-5 Diurnal measurements of OH reactivity illustrating the very high levels of atmospheric pollutants, particularly during the morning rush hour, at ground level in Mexico City, compared to U.S. cities and a rural site (data from W.H. Brune).

Aerosols affect climate both directly, by absorbing, reflecting, and scattering solar radiation, and indirectly, by influencing cloud optical properties, cloud water content, and cloud lifetime. The aerosol direct and indirect forcing may have offset as much as 50 to 75 percent of the greenhouse gas forcing since the Industrial Revolution times (NRC, 2005c). The climate influence of aerosols is one of the largest uncertainties in models of present and future climate. Furthermore, aerosols are a major component of air pollution with well-documented effects on human health, ecosystems, and visibility.

A major catalyst for advancing understanding of atmospheric aerosols was the INDian Ocean EXperiment (INDOEX), which culminated in a 1999 field campaign. The INDOEX campaign brought together researchers from the United States, Europe, India, and the Maldives for an intensive investigation into the factors controlling aerosols over the tropical Indian Ocean and the associated climate impacts. By integrating observations from satellites, aircraft, ships, surface stations, and balloons with one- and four-dimensional models, the participants made several striking discoveries. They observed remarkably high levels of aerosols extending over most of the South Asian region and the North Indian Ocean, and up to 3 km altitude (Ramanathan et al., 2001a). These aerosols enhance scattering and absorption of solar radiation, while also producing brighter clouds that are less efficient at releasing precipitation. Thus, solar irradiance that would otherwise reach Earth's surface is either reflected back to space or contributes to warming of the atmosphere directly, thereby changing the atmospheric temperature structure. Further, the aerosols were found to suppress rainfall, inhibit removal of pollutants from the atmosphere, and lead to a weaker hydrological cycle.

As the agency leading the mission, NSF was instrumental in planning and executing INDOEX. NSF program managers helped coordinate the contributions of DOE and NOAA, and worked with the U.S. State Department to coordinate the participation of other nations, especially India and the Maldives. Further, NSF supported and facilitated coordination among the many PIs involved to ensure proper utilization of resources.

In addition to NSF support for PIs and coordination efforts, INDOEX utilized two other modes of NSF support: the Center for Clouds, Chemistry, and Climate (C4) STC and NCAR/UCAR system. The C4 center was important for fostering INDOEX because it provided (1) a multi-institutional, multinational organization; (2) funding flexibility; (3) an established infrastructure, including a center manager; and (4) support for a testbed experiment preceding INDOEX to demonstrate capabilities in the field. Moreover, STC funds helped support the analysis part of INDOEX, whereas many field programs suffer in this regard.

UCAR played an important role in the field campaign logistics, particularly in the deployment of the aircraft, communication between platforms (i.e., ships, aircraft), and in providing meteorological forecasts in the field. NSF program managers led the design of this interagency program involving the participating PIs and UCAR in determining what instruments would be included and how best to deploy them. NCAR scientists brought to bear several modeling tools during the INDOEX campaign (e.g., chemical forecasts that helped guide flight plans) and to analyze the afterwards (e.g., aerosol assimilation models).

The INDOEX campaign combined with the focused research on aerosols at the C4 STC has had several significant scientific impacts. It has resulted in dozens of scientific papers, establishing a new paradigm for how air pollution and climate change are linked and stimulating a new area of interdisciplinary research. Furthermore, the field observations allowed for initial validation of how climate models treat aerosol forcing. Several follow-up research efforts are under way, including the establishment of

BOX 2-7
Serendipitous Path to Atmospheric Brown Clouds

Veerabhadran Ramanathan, Professor of Climate and Atmospheric Sciences
Scripps Oceanography Institute
Ph.D., Planetary Atmospheres, State University of New York at Stony Brook

Occasionally, scientists start an enquiry into a problem which leads to deviations from that path and ultimately to an unexpected discovery on a different problem. This is the case with respect to my work on atmospheric brown clouds (ABCs), NSF and its ATM played a major role in this serendipitous path.

After the 1975 discovery of the CFC's role as a super-strong greenhouse gas and its implications for similar effects by other manmade gases in the atmosphere, I became concerned about the human impact on climate. Subsequently in 1980, Roland Madden and I concluded that the global warming from greenhouse gases would manifest in the observed records by 2000. In order to get a glimpse into a future warm planet, my students and I began looking at one of the warmest oceanic regions, the western Pacific warm pool (WP2). Using the just-released Earth Radiation Budget Experiment (ERBE) data, we stumbled on the super greenhouse effect phenomenon, a mechanism for unstable warming. Yet the maximum surface temperatures over WP2 were remarkably stable on decadal time scales. That same year, with ERBE colleagues at NASA, I had shown that clouds had a large negative radiative forcing globally, which would produce a surface cooling. The two findings together with more data analyses led us to postulate that the negative shortwave cloud forcing would act like a thermostat, and maintain maximum temperatures in WP2 below 303 to 305 K in the absence of external forcing. This search for the Pacific Thermostat led to the discovery of the Indian Ocean ABCs, catalyzed by NSF's role discussed next.

In 1989, teaming up with Paul Crutzen, I proposed an NSF STC (the Center for Clouds, Chemistry, and Climate, C4) which was formed in 1991 with Jay Fein as the program monitor, who along with C4 associate director H. Nguyen played a major role in the subsequent developments. I proposed testing of the thermostat to Jay Fein who immediately realized its importance and put me in touch with Joachim Kuettner of NCAR. Together, we proposed the Central Equatorial Pacific Experiment (CEPEX) conducted in 1993. The infrastructure of C4 and NSF's help enabled us to mount a field campaign in a remarkably brief period of 1.5 years. The field data provided evidence for the key ingredients of the thermostat, but raised a major new gap in our understanding. We found out that the solar radiation reaching the sea surface was much lower than that predicted by models and concluded the difference was because of missing absorption processes within the atmosphere.

Stunned by this finding, I began focusing on the missing physics in models. It became clear that I had to account for absorbing soot aerosols. This realization led me to the North Indian Ocean [NIO] where aerosol effect is expected to be larger

because of its vicinity to populated southern Asia. Another reason was the challenge by Peter Webster that maximum temperatures in the NIO were not limited by the cloud thermostat. Paul Crutzen became interested in the NIO as well because of his long standing interest in tropical air pollution, and thus was born the Indian Ocean Experiment (INDOEX). INDOEX was a $25 million effort with NSF as the lead agency and multinational support (United States, Germany, Holland and India). Again, a program manager was crucial in nurturing INDOEX through NSF and other agencies in the United States and India.

The sobering finding of INDOEX was the discovery of a wide-spread brown haze and its large solar-dimming effect that masks global warming and impacts regional hydrological cycle. This discovery led to the United Nations Environmental Programme (UNEP)-sponsored ABC program (thanks to strong support from K. Toepfer and S. Shresta of UNEP), with participating scientists and governmental institutions from China, Germany, Japan, India, Korea, Sweden, and the United States, to study the combined effects of global warming and wide-spread Asian air pollution on regional and global climate, water budget, agriculture, and health.

numerous observatories in the Indo-Asia-Pacific region to monitor aerosol pollution and a UNEP-sponsored Project Atmospheric Brown Cloud.

The identification of the importance of tropospheric aerosols is a particularly good example of how small centers, such as the C4 STC, can foster major scientific breakthroughs and have the flexibility to respond to unforeseen avenues of research. Further, it illustrates how ATM successfully facilitated international cooperation through the use of large national centers and domestic interagency coordination. Support for numerous types of activities played important roles in this research, including technology development, field programs, and laboratory research.

Case Study 9: The Role of Mauna Loa Measurements in Understanding the Global Carbon Cycle

In 1958, Charles Keeling began making measurements of atmospheric carbon dioxide (CO_2) at the Mauna Loa Observatory. The long-term monitoring record from this site has become an icon of global climate change. Keeling's first measurements at the Mauna Loa Observatory began in 1958 and were funded by Dr. Henry Wexler of the U.S. Weather Bureau as part of the bureau's efforts during the first International Geophysical Year to measure CO_2 at remote locations. The possibility of making continuous measurements of atmospheric CO_2, built on improvements in instrumentation using infrared gas analyzers and Keeling's carefully developed manometric technique for precisely calibrating analyzers. Keeling's interest in atmospheric carbon dioxide measurements was fueled by his earlier observations of the variability in CO_2 concentrations near the ground in Pasadena and Big Sur State Park in California, the Olympic Peninsula, in Washington, and the high mountains of Arizona (Keeling, 1958) as well as insights from reading *Climate Near the Ground* (Geiger, 1957). It is also interesting to note that Keeling worked with Sam Epstein to make ^{13}C isotopic measurements of his first samples, foretelling the importance of both atmospheric carbon dioxide and the accompanying carbon isotopes in understanding the global carbon cycle almost 50 years after the first Mauna Loa measurements (Keeling, 1958).

Once Dr. Keeling had secured funding for his atmospheric carbon dioxide measurements, getting the measurements set up required his attention to competing demands (Keeling, 1998). The backing from the U.S. Weather Bureau allowed Keeling to purchase four instruments—one for deployment at Mauna Loa, one for deployment in Antarctica at the Little America Station, one for deployment on the Scripp's Oceanographic Institute research ship, and a fourth for use in the laboratory to cross calibrate with Keeling's laborious manometric technique. Dr. Keeling, director of Scripps Oceanographic Institute, succeeded in attracting Dr. Keeling to

Scripps to pursue his CO_2 studies there. Despite the difficulty of making measurements in Antarctica, it became the first monitoring site established in 1957. Dr. Revelle was convinced that shipboard and aircraft measurements were critical to understanding what was then thought to be the substantial spatial variability of CO_2. Revelle's interests competed with Keeling's determination to set up the Mauna Loa site. The U.S. Weather Bureau was able to provide a full-time employee to assist Keeling in meeting these competing demands and Keeling began the Mauna Loa measurements in 1958.

While the Mauna Loa data record is remarkably continuous over almost 50 years, the continuity of the funding record required the participation of a number of funding agencies and backers, including the international community. The publication of the first description of the seasonal cycle of CO_2 in 1960 was crucial for building support for his efforts (Keeling, 1960). The Mauna Loa and South Pole measurements were supported through 1962 by funding from the NSF and the U.S. Weather Bureau. In 1963, the Weather Bureau funding was discontinued and the entire program was funded by NSF. In 1961 and 1969, Keeling went to Europe on sabbatical, which was important for stimulating international interest in these measurements and for the establishment of Keeling's laboratory as the central laboratory for the WMO's CO_2 calibration effort. In 1971, NOAA installed an additional CO_2 analyzer at Mauna Loa. Also in 1971, NSF reduced funding for Keeling's laboratory by 50 percent, deeming that atmospheric CO_2 measurements were routine. The head of the WMO, Dr. Christian Junge, helped to restore NSF funding and worked to secure funding from the newly formed UNEP to calibrate CO_2 measurements worldwide based on the scientific arguments laid out in Keeling's 1970 paper (Keeling, 1970).

The next big scientific and funding breakthroughs came with Keeling's publication of 14 years worth of CO_2 measurements from both Mauna Loa and the South Pole (Keeling et al., 1976). That same year, the Atomic Energy Commission established the Energy Research and Development Agency (ERDA) to argue for nuclear-powered electricity generation. ERDA pursued studies of the carbon cycle because "the burning of fossil fuels might be more dangerous to mankind than any perceived side effects of nuclear energy" (Keeling, 1998, p. 56). ERDA became the DOE and the precedent was set for DOE to pursue studies of the carbon cycle and fund the Mauna Loa CO_2 measurements in 1978. This occurred just as the NSF was decreasing its support for Keeling's measurements because they were again considered to be routine. In 1980–1982, NOAA and DOE contributed 80 percent of the funding for the Mauna Loa measurements and NSF funding was slated to be eliminated at the end of 1982.

More than two decades of measurements at Mauna Loa laid the groundwork for what would be the next big breakthroughs in the carbon cycle: the return to the measurement of carbon isotopes; the measurement of CO_2 in ice cores; the development of models of the carbon cycle with north–south resolution of sources and sinks of atmospheric carbon; and, eventually, the three-dimensional representations of the global carbon cycle (e.g., Box 2-8; Fung, 1986; Heiman and Keeling, 1986; Fung et al., 1987; Keeling et al., 1989). However, continual funding for Mauna Loa would be problematic. In 1981, NOAA's responsibility for funding of Mauna Loa measurements was transferred to DOE. In 1983, DOE indicated that the agency was withdrawing funding. Once more, Keeling successfully reapplied to NSF for support of the Mauna Loa measurements and the WMO CO_2 program calibration. NOAA funding was secured once more, but only for one year. After considerable argument and discussion in 1984, DOE began to fund the Mauna Loa CO_2 measurements once again and that funding would continue through 1994. In the funding hiatus of 1982, the Electrical Power and Research Institute began to support the Mauna Loa CO_2 measurements as well.

The challenges in maintaining funding for continuous measurements of CO_2 at Mauna Loa continued until Keeling's death. The history of funding for the Mauna Loa record underscores the resilience rooted in maintaining healthy relationships with multiple funding agencies, the importance of establishing and maintaining international partnerships, and effective interdisciplinary collaboration within the scientific community. Charles Keeling's persistence and passion for his subject are a testimony to the difference a single individual can make. For its part, ATM was able to ensure that support for individual PIs, facilities, and instruments continued, by adapting to the changing contributions and priorities of other agencies.

Case Study 10: Improving El Niño Predictions

Jacob Bjerknes, in a series of papers between the late 1960s and mid 1970s, laid out important groundwork for understanding El Niño, or El Niño/Southern Oscillation (ENSO) as a coupled atmosphere–ocean phenomenon. The work of several others following him (e.g., Wyrtki, 1975; McCreary, 1976; Busalacchi and O'Brien, 1981; Cane, 1984) elaborated the manner in which the tropical Pacific Ocean responds to changing patterns of wind during El Niño events. In parallel, a number of modeling studies reinforced Bjerknes' conclusions concerning the influence of sea surface temperature anomalies on the tropical and extratropical atmosphere associated with El Niño (e.g., Rowntree, 1972; Lau, 1981; Zebiak, 1982; Shukla and Wallace, 1983). The seminal work by Rasmussen and Carpenter (1982) provided a coherent description of the systematic evolution of oceanic and atmospheric anomalies during El Niño events, as derived from collected historical observations over several decades.

The outsized El Niño event of 1982, with its similarly outsized impacts felt throughout much of the globe, did much to galvanize global attention, and to redouble the research community's resolve to better understand the phenomenon and its potential predictability. Through this one event, the importance of ENSO to societies worldwide came into much sharper focus.

Following a very few pioneering efforts to model atmosphere–ocean interactions underlying ENSO (e.g., McWilliams and Gent, 1978; Lau, 1981; McCreary, 1983), a number of new studies were undertaken in the mid 1980s to provide a more comprehensive understanding of ENSO dynamics and associated predictability. One line of research, introduced by Barnett (1981, 1984) and later by Graham et al. (1987), applied advanced statistical methods to identify systematic lead-lag relationships between atmospheric and oceanic variables, and to exploit them to develop statistical prediction models for El Niño related sea surface temperature patterns. This work indicated real predictability associated with El Niño onset, at lead times of several months.

The second line of research involved further development of physically based models. Several were developed, with various simplifying assumptions (e.g., Anderson and McCreary, 1985; Cane et al., 1986; Schopf and Suarez, 1987). The approach taken by Cane and Zebiak (Cane et al., 1986; Zebiak and Cane, 1987) proved particularly useful and led to the first dynamical El Niño predictions. The most important simplification in this case was to model only the departures of atmospheric and oceanic states, relative to the observed, seasonally varying mean climatological state. In so doing, a relatively simple dynamical model was capable of simulating realistic features of El Niño, including aperiodic oscillations with a spectral peak near the four-year period, and the characteristic spatial patterns and magnitude of anomalies. These authors introduced an El Niño forecasting system based on this model in 1986, and at that time also produced a one-year lead forecast indicating that a moderate-amplitude El Niño event would develop later that year. The forecast proved substantially correct, though the onset was more than two months later in nature. Routine predictions with this system were initiated the following year, and have continued (with several revisions) to present.

The late 1980s were a period of great excitement in the climate community. A multitude of El Niño prediction systems (both statistical and dynamical) were developed and routine predictions of El Niño were established with several of these systems. Several of the more successful models were analyzed in some detail to understand better the dynamical basis for the empirically observed predictive skill. A major new, near-real-time observing system was designed and substantially deployed to monitor the upper ocean and surface ocean–atmosphere conditions in the equatorial Pacific. Its purpose was to provide more detailed study of ENSO physics, and to supply

BOX 2-8
Carbon Cycle Research

Inez Fung, Co-Director
Berkeley Institute of the Environment
Sc.D., Massachusetts Institute of Technology, Meteorology

Guided by an extremely stimulating but often-absent mentor, Jule Charney, I learned my Doctor of Science (Sc.D.) in Meteorology from the Massachusetts Institute of Technology. My dissertation was on the organization of spiral rainbands—instability of a vortex flow with shear in two directions. I had learned Fortran in anticipation of a modeling thesis, but Charney insisted that I solve my problem analytically, as I would be working with models for the rest of my career. After graduation, I became a National Research Council post-doc at NASA's Goddard Space Flight Center in Maryland and then a research associate at the Lamont Doherty Geological Observatory of Columbia University and NASA's Goddard Institute for Space Studies (GISS) in New York City.

My relocation from Maryland to New York, and my scientific move from geophysical fluid dynamics to the carbon cycle was at the suggestion and encouragement of Charney, who instructed me not to confuse my work with my life, and told me to join my husband, who was a post-doc at Lamont Doherty. It was at GISS under Jim Hansen's leadership that the first three-dimensional global carbon model came together, with the new atmospheric tracer transport model developed by Gary Russell at the core. This work at NASA eventually led to the locating of the missing carbon dioxide in the atmosphere (surprise, it is under our feet!) and the role that the Northern Hemisphere's terrestrial biosphere plays in taking up some anthropogenic carbon dioxide. During that period, NASA funded Interdisciplinary Science teams to support the Earth Observing System, with a typical funding period of 10 years. I was a co-investigator on the proposal led by Piers Sellers to study biosphere–atmosphere interactions using global models ("top-down view"). At the insistence of NASA Headquarters, the Sellers (east-coast) team merged with Harold Mooney's (west-coast) team, which provided a "bottom-up view" of the same interactions. As a result, I learned plant physiology, biogeochemistry, and remote-sensing science from the other team members. It was a great time of learning and friendship. Together, we combined the major pieces to model atmosphere-biosphere exchanges of energy, water, and carbon at the global scale, with the models going from stomatal conductance and microbial respiration to global climate change.

After 16 years with NASA, I moved to the University of Victoria in Canada for 5 years and then to the University of California at Berkeley where I became the Director of the new Berkeley Atmospheric Sciences Center. While I was with NASA and in Canada, I could not apply for or receive NSF funds. And so it was some 20 years after my doctorate that I wrote my first NSF proposal. The proposal was to couple terrestrial and oceanic carbon cycles to the NCAR Community Climate System Model (CCSM) and apply the resulting model to study carbon-climate feedbacks. The idea grew out of several events that happened at about the same time. The first was the joint meeting of the WCRP-WGCM and IGBP-GAIM groups in Melbourne, Australia, in 1998. The WCRP-WGCM group had just finished their 1 percent/yr CO_2 experiments with coupled atmosphere-land-ocean-ice climate models and were ready for the next challenge. After much back and forth, I proposed that we (IGBP-GAIM) could replace their specification of CO_2 forcing in the climate models with CO_2 predicted as a result of specified fossil fuel emissions and model-calculated land and ocean carbon exchanges.

Finally, there were smiles. I nicknamed the experiments "the Flying Leap Experiments" as we were not progressing systematically and were bound to "go splat." The second event was the formation of the Biogeochemistry Working Group (BGC WG) in the NCAR CCSM framework. Scott Doney and I were the first co-chairs of the BGC WG, and we started the "leap series" for the CCSM to include interactive biogeochemistry to the CCSM.

NSF proposal submissions require a report on work supported by prior NSF grants. I was afraid that the reader might interpret "none" to mean "no work" rather than "no NSF grant," and so I included summaries of all my previous work. Because the proposed work involved a physical climate model coupled to terrestrial and ocean biogeochemistry, which was totally new at the time, it was difficult to include the background science, identify the scientific need, and lay out a research strategy in 15 pages. The proposal took a prolonged period to review, and was finally handled by three NSF program managers (Atmospheric Chemistry, Climate Dynamics, and Oceanography). NSF support for the work is greater than the grant to Berkeley, as the proposed work built on and had the support of the CCSM team at NCAR, and had access to the NCAR computers. I am grateful to NSF program managers for recognizing the importance of the endeavor, and their patience in piecing together the support from the different programs. I am pleased that as a result of the Melbourne meeting in 1998 and this research, there are now over ten international groups with prognostic CO_2 in their climate models.

It is fair to say that much of the most interesting research I have done was started with conversations and vague ideas that would not have fit into routine program portfolios in any funding agency and would not have survived peer review. Incubation of the ideas was luckily made possible by piggybacking on other grants and by understanding superiors and program managers.

necessary information to the predictive models (McPhaden and Hayes, 1990). And, in order to study coupled processes in the so-called Warm Pool region of the western tropical Pacific—processes poorly captured by existing models and believed to be a limiting factor in prediction—the proposal was put forward to undertake a major observational field campaign: the Coupled Ocean Atmosphere Response Experiment (COARE; Godfrey et al. [1998] document the many accomplishments and findings of this major program). Finally, the international research community developed the concept of an International Research Institute, which would transition El Niño and seasonal climate predictions into operational forecasts, and address the myriad issues at the interface of climate and societal needs, allowing the perceived benefits of this new climate knowledge and information to be realized in practice. All of these activities took place within the context of the international TOGA research program undertaken by the World Climate Research Program during the period 1985-1995. The U.S. participation in this program was organized through an interagency process, and included major contributions from NOAA, NSF (ATM and OCE), and NASA.

The key research that laid the foundation for the first efforts at El Niño prediction was undertaken through grants to individual or small groups of PIs, and was supported by NOAA, NSF (ATM and OCE), and NASA. This initial research was the formative work in understanding the coupled processes and how to forecast El Niño; it opened the door to operational prediction and its multiple societal benefits. The basic investment in research early on catalyzed investments in the observing systems and modeling efforts by other agencies. Most notably, NOAA advanced two extremely important programs: the Tropical Atmosphere Ocean observing system, and the coordinated modeling and prediction program known as the TOGA program on prediction (PIs were still supported through individual grants). UCAR provided access to facilities and field support throughout much of these efforts. Finally, the major field campaign TOGA COARE was international in scope, but very substantially supported by NSF/ATM, NSF/OCE, NOAA, and NASA. The U.S. interagency coordination in all of these efforts was outstanding and highly effective.

Case Study 11: Development of Helioseismology

Observations of the solar surface reveal oscillations of the order of five minutes (Harvey, 1995). These surface oscillations, a manifestation of resonant oscillations within the Sun, provide a window into the Sun's interior structure and dynamic that has not only led to a golden age in solar physics and stellar physics, but it has also led to fundamental new understanding in atomic physics. There are about 10^7 pressure modes oscillating in the Sun with typical amplitudes of ~1 cm s⁻¹ (Harvey et al., 1996). Through a complex but straightforward multistep process, sequences of images of the Sun are isolated into the Sun's normal modes, which can then be used to obtain information about the structure and circulation of the Solar interior.

The basic ideas for helioseismology came from the university community and one of the Astronomy federally funded research and development centers (Harvey, 1995). In 1960, Robert Leighton of Caltech discovered that the surface of the Sun was characterized by small-scale patterns that oscillated radially with a period of about five minutes. In the early 1970s, Roger Ulrich of UCLA and John Leibacher of the National Solar Observatory (NSO) and Robert Stein of Michigan State University demonstrated that the oscillations were caused by acoustic waves generated in the solar interior—the Sun "rings like a bell." In 1975, Franz Deubner observed the predicted patterns and used it to demonstrate that the then current solar models were incorrect.

Obtaining the surface oscillations is a challenge because of their small amplitudes (Harvey, 1995; Gough et al., 1996; Harvey et al., 1996). Furthermore, longer time series—significantly longer than the consecutive hours of daylight in low latitudes—are needed to determine the normal modes from the portion of the Sun visible from Earth than if the whole surface could be sampled. Thus, the early measurements were made at the South Pole supported by NSF Polar Programs and by a "relay" involving several observatories around the world. These early measurements, however, could only sample those modes with wavelengths of the order of the diameter of the Sun.

The next step, obtaining long-term high-resolution observations of the Sun's surface, involved two major projects: the Global Oscillation Network Group (GONG), sponsored by NSF/Antarctic Submillimeter Telescope (AST) and run by NSO, and The SOlar and Heliospheric Observatory (SOHO), sponsored by the European Space Agency (ESA) and NASA. A competition was held for the best design for the GONG instruments; the winning design, the Fourier tachometer, was developed by Timothy Brown of HAO/NCAR along with Jack Evans and others at NSO.

The GONG network of six identical solar-imaging telescopes has been collecting continuous data of the Sun's surface since 1995; the GONG Advisory Panel included scientists from NCAR and a number of universities. Some complementary efforts aimed at getting longer temporal baselines are the U.K. Birmingham Solar Oscillation Network; while efforts aimed at getting structure deeper in the Sun include the Mount Wilson-Crimean-Kazakhstan mini-network, the high-degree helioseismometer operated by the NSO at Kitt Peak, and the low-and intermediate-degree experiment (LOWL), operated at NCAR's HAO at Mauna Loa Observatory. Scientists from several U.S. universities are involved in helioseismology, as are scientists in Australia, Denmark, France, Germany, India, Japan, Taiwan, and the United Kingdom. The Solar Terrestrial Research Program in ATM provides an average of roughly $400K–500K per year for helioseismology through its regular grants program, and through the National Space Weather Program (NSWP), Research at Undergraduate Institutions, and NSF faculty early career development program (CAREER) awards.

In its first decade, helioseismology has ushered in a golden age in solar research. Solar structure and motions have been clarified. One of the first results from helioseismology was that the convection zone extends downward to 0.713 of the solar radius, significantly deeper than many earlier solar structure models had predicted (Harvey, 1995). Helioseismology has also demonstrated that the zonal flow at the surface (long documented by following the trajectories of sunspots and other surface features) changes little with radius through the convection zone; but there is considerable radial shear between the convection zone and the radiative interior, through a layer called the "tachocline." Helioseismic measurements have also shown that small variations in solar rotation—so-called torsional oscillations—occur throughout the convection zone, with periods tied to the solar cycle period. Beyond differential rotation, helioseismic inversions have revealed several other motions, including poleward meridional below the photosphere and other "solar weather" patterns near the surface. Inversions are also beginning to reveal the thermal structure below "active regions," the sites of sunspots. Now solar activity on the side of the Sun opposite the Earth is routinely detected by so-called "far-side imaging."

These findings inspired a new generation of individuals and groups around the world to develop idealized models and full-blown general circulation models of the outer layers of the Sun, designed to explain the solar dynamo and the solar cycle.

In 2006, Mausumi Dikpati, Giuliana de Toma, and Peter Gilman, of NCAR's HAO used a so-called "flux-transport" dynamo model that assimilated data from the previous three solar cycles (borrowing a technique from numerical weather prediction) to predict successfully the relative amplitudes of the new cycle for each of the last nine solar cycles, and to project the amplitude of the next solar cycle in 2012 (Figure 2-6) (e.g., Box 2-9; Dikpati et al., 2006). These predictions and future results coming from helioseismology-inspired research will provide significant contributions to predictions of Space Weather.

In addition, helioseismology has offered a new way to constrain the distribution of elements making up the Sun. The elemental abundance has historically been determined by using what we know—luminosity measurements, solar radius, and solar spectral results giving clues to the chemical composition—to build one-dimensional solar and stellar structure models. More recently, helioseismology has been used to determine the sound

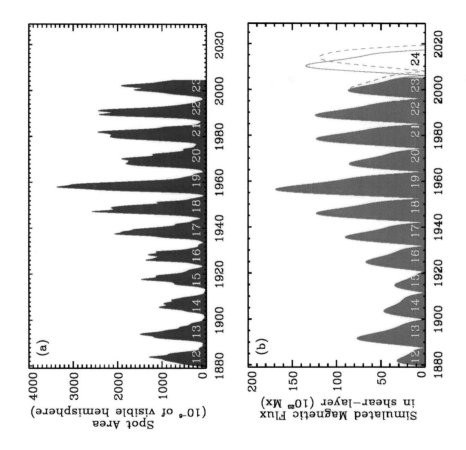

FIGURE 2-6 (a) Observed sunspot area, smoothed by ~1-year Gaussian running average, plotted as a function of time; (b) simulated toroidal (zonal) magnetic flux at the bottom of the solar convection zone, which is the source of sunspots. Solid red area and curve are for steady meridional flow; dashed red curve is for time-varying flow since 1996 incorporated. SOURCE: Dikpati et al. (2006).

speed, providing an independent route to constrain the composition. Until recently, both approaches were consistent with the same solar abundance of elements—until Australian Martin Asplund and colleagues (Asplund et al., 2005) used spectral data interpreted with the help of three-dimensional simulations to revise the solar chemical composition to reduce the percentage of atoms heavier than helium. Sarbani Basu of Yale University, funded by a CAREER grant through NSF/ATM/Solar Terrestrial Research, was

BOX 2-9
Predicting Solar Cycles

Mausumi Dikpati, Scientist
NCAR
Ph.D., Physics, Indian Institute of Science, Bangalore, India

After successfully completing my Ph.D. in India, I decided to come to the United States for my postdoctoral study. I chose the United States because I thought it would be the best place to pursue my area of research, namely the modeling and predicting of solar cycles. I was accepted in the Advanced Study Program of NCAR for a post-doc in 1996 to work with a group of scientists that work in this same area of research. I stayed in this position until 1999, during which time I not only gained much experience in solar cycle modeling but I was also fortunate enough to get the chance to work with Peter Gilman on a new topic, instabilities in the solar tachocline. Interactions with him on professional matters and otherwise have been invaluable in shaping my career and personality.

My accomplishments during this post-doc provided me the opportunity to continue my research at NCAR as a NASA-funded project scientist until 2002. During this time I worked on symmetry selection in solar cycle dynamo models. This research led to a paper that was nominated for the UCAR outstanding publication award. These further accomplishments helped me win the NCAR-wide competition for a new NSF base-funded scientist I positions in 2003. I was subsequently promoted to scientist II in 2006.

Apart from science, I am very involved in spiritual activities that provide me more focus and insight. These activities also brought up whether I could do research that would benefit society as a whole, for example, whether I could develop a model to predict solar cycles. I realized that I might be able to do that by building a predictive tool from my research on so-called "flux-transport" dynamos applied to the Sun. I actively started the work and, fortunately, found the right colleagues—Charles Nick Arge (AFGL), Paul Charbonneau (University of Montreal), Giuliana de Toma (HAO/NCAR), David Hathaway (NASA/MSFC), Keith MacGregor (HAO/NCAR), Matthias Rempel (HAO/NCAR), and Dick White (HAO/NCAR)—to work with. Both the NSF and NASA have supported this research.

I continued publishing papers in the *Astrophysical Journal* with results from the development of the predictive tool. Feature articles, discussing my work in popular

on a team that found a potential way out of this impasse. By raising the abundance of neon (the most uncertain of the heavier element abundances), they have shown that the rest of the revised abundances are potentially consistent with both helioseismic and spectral results (Bahcall et al., 2006). She and H.M. Antia are now attempting to measure the total solar heavy element abundance through helioseismology (Antia and Basu, 2006). This

magazines such as *New Scientist* and *National Geographic* in 2004, have given me great encouragement that this research is of interest and value to society. After our prediction for the next solar cycle was published in *Geophysical Research Letters* on March 6, 2006, the work received great attention in the worldwide press. When reports were translated into Bengali and reported in "Songbad Protidin" (the daily newspaper in Calcutta), it caused my mother to become very excited about the recognition her daughter was getting.

is important because stellar models and models of the universe are tied to solar abundance models.

The reach of helioseismology extends beyond solar physics to particle physics. Secondary students all over the world are taught that the Sun is powered by a reaction that converts hydrogen to helium, releasing neutrinos in the process. In the 1960s, Raymond Davis of Brookhaven

National Laboratory (University of Pennsylvania after 1984) set up an experiment to detect solar neutrinos, 4,800 feet below the surface in the Homestake Gold Mine, in Lead, South Dakota. Neutrinos were detected, but only one-third the amount predicted assuming that the commonly accepted helium-to-hydrogen conversion was true; stimulating new detectors around the world—Kamiokande in Japan, SAGE in the former Soviet Union, GALLEX in Italy, and Super Kamiokande (*http://nobelprize.org/physics/laureates/2002/davis-autobio.html*). Those explaining the "missing" neutrinos invoked a rapidly rotating solar core, contrary to the thinking of the solar physics community. The LOWL instrument revealed a core rotating at a rate similar to the outer layers of the Sun, indicating something else was needed to explain the missing neutrinos. Also, John Bahcall, Sarbani Basu, and Marc Pinsonneault argued that the fact that the sound speed and density in the core of standard solar models are so close to those inferred from helioseismic measurements, also implies that explaining the missing neutrinos would mean invoking nonstandard neutrino physics rather than nonstandard solar models. Finally, in 2001–2002, scientists at the Sudbury Neutrino Observatory in Ontario, Canada, found evidence that the neutrinos could oscillate among three forms. The 2002 Nobel Prize in Physics was awarded to Davis and Masatoshi Koshiba of the University of Tokyo "for pioneering contributions to astrophysics, in particular for the detection of cosmic neutrinos."

Responsibility for funding helioseismic studies has been shared at NSF between ATM and AST, and within the United States has been shared between NSF and NASA. The partnership between NSF's ATM and AST divisions exemplifies a successful intra-agency partnership across directorates within NSF. The helioseismic observations are made both from space and from worldwide ground-based networks, and analysis of the data is carried out at many U.S. and foreign universities who meet regularly to discuss the latest results. The ground-based GONG network has recently been upgraded to provide much higher spatial resolution with NSF/AST money and within two years the SOHO satellite will be superceded by the new Solar Dynamics Observatory satellite (NASA Living with a Star Program, ESA). NSF/ATM and NASA continue to support helioseismic studies with grants to university PIs.

Case Study 12: Reading the Paleoclimate Record

Given the relatively short instrumental climate data record, the ability to test climate models depends strongly on the availability of paleoclimate records. Advances in paleostudies over the past few decades have significantly extended the data record, providing a context for the instrumental climate record and the evolution of atmospheric composition (e.g., CO_2,

CH_4, NO_2). Paleorecords are used to infer the impact of anthropogenic contributions on long-term climate trends and to test climate models, particularly regarding the sensitivity of climate to CO_2.

A major milestone in the field of paleoclimatology was the demonstration that the concentrations of oxygen-18 and deuterium accumulated in deep cores of glacier ice can be used as an indicator of past temperature (Dansgaard, 1964). Radioisotopic dating methods, such as the radiocarbon dating methods developed in the early 1970s for dating sediment or lake cores, were applied to these data sources in order to construct consistent chronologies of past temperature. Since that time, many other radioisotopic dating methods have been tailored and calibrated for specific application to paleoclimate datasets (Cronin, 1999).

The first ice cores were obtained from Vostok, Antarctica, in the 1970s by Russian scientists. In the 1980s, French and American scientists subsequently joined this effort, which was supported primarily by NSF's OPP (Box 2-10). The most recent Vostok drilling yielded the longest record, dating to 420,000 years B.P. (Petit et al., 1999). Similar drilling expeditions to Greenland have yielded data records on atmospheric composition and temperature trends on millennial time scale. These records have revealed the climate's variability over millenial cycles, and have provided evidence for the correlation between global temperatures and CO_2 concentrations. These records also demonstrate the importance of Milankovitch cycles in regulating climate (Imbrie and Imbrie, 1979; Berger et al., 1984).

Microfossil records from ocean cores led to the discovery of past systematic changes in sea surface temperatures and changes in the amount of glacial ice stored on the continents. Such records also permit the reconstruction of past global ocean currents, which is important because of the influence of the global thermohaline circulations on climate. Some of these breakthrough discoveries were made during the large CLIMAP (CLimate: Mapping, Analysis, and Prediction) program, a multi-institutional consortium effort funded by the NSF and led by J. Imbrie, J.D. Hays, N. Shackelton, and A. McIntyre. This effort led to a follow-up called COHMAP, which was supported primarily by Climate Dynamics Program of NSF and by DOE (carbon dioxide research division). An important and surprising finding was revealed by the Greenland Ice Core Project, where scientists detected the ability of abrupt climate shifts (5–10°C) during an interglacial period (Dansgaard et al., 1989). For centennial, decadal, or even year-to-year resolution in past climate variability, paleo proxies such as tree-rings, coral records, and lake or bog sediments are used (NRC, 2002a, 2006b).

Many of the field-intensive drilling expeditions, such as ocean sediment or glacial ice core drilling, were supported by international efforts and funded by multiple agencies such as NASA, NOAA, NSF, and DOE. While OPP plays the principal role in funding ice core drilling operations and the

BOX 2-10
Reconstruction of the Earth's Paleoclimate from the Highest Mountain Glaciers

Lonnie Thompson, Professor,
Ohio State University
Ph.D., Geology, Ohio State University

In 1973, when I first started thinking about drilling ice cores on tropical mountains as a graduate student at Ohio State University (OSU), it was not possible to get NSF funding for ice core retrieval outside of the polar regions. At that time drilling into high-altitude tropical glaciers was considered technically unfeasible and scientifically mis-guided. When I began to study the Quelccaya ice cap in southern Peru, I was funded by Jay Zwally at NSF's OPP from $7,000 left in the budget after all the other projects had been supported. This research produced a record of tropical climate from the snow pit and shallow core studies we retrieved from Quelccaya.

In 1978 our proposal to drill through Quelccaya was accepted by NSF's new Office of Climate Dynamics (OCD). After an attempt to transport a conventional drill up the ice cap failed, we had the idea of developing a light-weight solar-powered drill that could be back-packed in pieces up the mountain. We attempted to convince OCD and the ATM to fund this wild idea, but were dealt a serious setback when one of the reviewers, who was a pioneer in polar ice core drilling, told NSF that he believed Quelccaya was too high for humans to live long enough to achieve this objective. The reviewer went on to remark that the technology did not exist to develop such a drill. However, a new program director at ATM, Hassan Virji, gave us an opportunity to test the frontiers of ice core drilling by funding our proposal, while OPP funded the drill development. In 1983 the OSU team and equipment made it to the top of Quelccaya and we drilled the first two ice cores to bedrock from a tropical ice cap.

From this precarious start, the Ice Core Paleoclimate Research Group developed at the Byrd Polar Research Center. Since that time we have successfully completed 50 such expeditions with the continued support of NSF, particularly ATM's Paleoclimate Program and OPP. This year we will conduct a cooperative ice core research program

analysis of such ice cores, NSF's ATM supports most other paleoclimate studies, such as tree-ring, lake sediments, and coral paleoclimate studies. In fact, ATM has consistently led in the support of tree-ring research (e.g., Figure 2-7).

The paleoclimate program is in a unique position because, although about 25 percent of all proposals submitted to ATM are in this area, the program has one of the smaller budgets in this division. However, the pro-gram is successful because the cross-division collaborations within GEO and NSF reflect the strongly interdisciplinary nature of this area of scientific research.

in the southwestern Himalayas near the source of the Ganges and Indus Rivers, made possible in part through funding from NSF's ATM-ESH program. We have had many dedicated program mangers over the years who promoted the paleoclimate community and their faith in this effort has produced a rich record of paleoclimate in regions where it once was unknown.

ANALYSIS OF THE CASE STUDIES

We turn now to the following questions: What light do these major accomplishments shed on the role of NSF ATM in its support of atmospheric sciences? In particular: What do they imply about the balance between the various modes of support, whether that balance has in fact been adjusted over time, and whether they provide evidence that there is a need to alter the balance?

The first observation is that NSF ATM has played a role in every one of these major accomplishments. In a few cases, ATM played only a minor

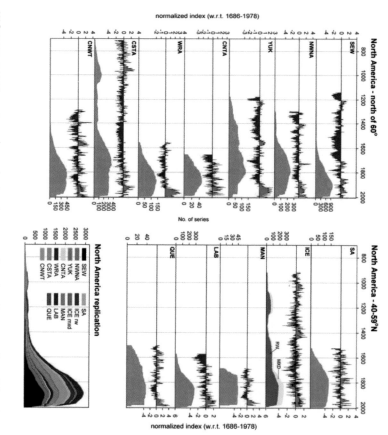

FIGURE 2-7 Tree-ring chronologies for several regional composites. The time series have been loosely grouped according to latitude bands and normalized to a common period. The bottom two panels in the right column show grouped replication plots for both North America and Eurasia. NOTE: ALPS = Alps, CNTA = Central Alaska, CNWT = Central Northwest Territory, CSTA = Coastal Alaska, ICE = Icefields, JAEM = Jaemtland, LAB = Labrador, MAN = Manitoba, MON = Mongolia, NWNA = Northwest North Alaska, POL = Polar Urals, QUE = Quebec, SA = Southern Alaska, SEW = Seward, TAY = Taymir, TORN = Tornetraesk, WRA = Wrangells, YAK = Yaktutia, YUK = Yukon. SOURCE: D'Arrigo et al. (2006). Reproduced by permission of American Geophysical Union; copyright 2006.

or supporting role but in the majority of these cases, NSF ATM's role has been central. Furthermore, the case studies demonstrate that all the modes of support—PI grants, including those for exploratory projects and in response to focused solicitations; small centers; the large national center; cooperative observing facilities; and field programs—have been important to one or more of these major science achievements. Likewise, each major achievement benefited from several modes. For example, much of the early work in climate modeling was supported by individual PI grants, however, with increasing complexity and the need for ever larger computing power, interagency and intersector support became increasingly important and the available computing facilities at NCAR ever more central to running the models. It is also difficult to envision that the space weather research community would have made the advances without the combination of modes such as small PI grants, support from NCAR for model runs, observing facilities, ATM initiatives, and the pioneering work supported by the small center.

Thus, the range of available modes has been a tremendous and necessary asset for the atmospheric sciences. This is a reflection of the nature of atmospheric science and its development over the past decades as discussed in detail in Chapter 3; but it is also evident that ATM's portfolio of modes of support and the balance among the modes has evolved with the state of the science. For example, at a very early stage, severe weather research was an interagency effort, mostly between NOAA and NSF's ATM, and within ATM supported by individual grants to University scientists, who worked in close collaboration with NCAR scientists. As it became necessary to integrate field observations, modeling capability and instrument development to advance severe weather research, the field was ripe to take advantage of new modes such as NSF's STC leading to the development of the world's first storm-scale prediction system. Another example is carbon cycle research, which began with a single PI effort originally funded by the U.S. Weather Bureau. Eventually, it became an interagency, cross-disciplinary, and multimode effort to support direct CO_2 measurements, ice core measurements, and the development of carbon cycle models.

Grants to individuals and teams of PIs were instrumental in all of the achievements, while the large national center contributed to nearly all of them. This reflects in part the fact that these have been the two dominant modes of support utilized over the past 40 years. But, more importantly, it reflects that these two modes have been effective at fostering a productive research environment. In addition, in more than half of the major case studies, the science was significantly advanced by field programs, often large efforts requiring significant coordination among researchers, different agencies, and in many cases different nations. The U.S. participation and interagency coordination during TOGA, an effort to further the understanding of ocean-atmosphere processes related to ENSO, exemplifies the success and importance of international and interagency field campaigns in advancing atmospheric research.

Some of the newer modes, such as small centers and cooperative observing facilities, have not been available as long and are a smaller portion of the ATM funding portfolio. Even so, the three small centers that have been established in the atmospheric sciences (CAPS, C4, and CISM) have each been engaged in research that either led to a significant leap in understanding, as in the case of CAPS and C4, or else are helping us to bring to fruition a major achievement, as in the case of CISM. Another advantage of these small centers is the explicit role for technology transfer. Indeed, atmospheric science is special in that one of the key transfer targets is the federal government.

The value of partnerships with other disciplines, agencies, and nations is also apparent in reviewing these case studies. Every major achievement analyzed required coordination with other agencies, including NOAA, NASA, DOE, EPA, and the Department of Defense. In some cases, broad interagency programs like the U.S. Global Change Research Program or the NSWP have played an important role in focusing research objectives and applying the collective resources of several agencies. Given the range of partnerships employed in these case studies, it is fair to conclude that NSF has been effective in fostering collaboration.

An important lesson to be gleaned from the research activities leading to these major accomplishments is that ATM has adjusted the balance from time to time as opportunities, needs, and scientific progress made necessary and possible. For example, when it became apparent that a concerted, coordinated effort could lead to significant advances in space weather prediction, ATM supported members of the scientific community in their bid for an STC, resulting in the recently formed CISM. The creation of the interagency U.S. Global Change Research Program in the late 1980s is another example of NSF ATM, in coordination with other agencies, identifying the need for greater organization and coordination, and then taking the steps to address this need. In general, ATM has been responsive to evolving needs and has effectively interacted with the community in choosing new directions. It does not in any way detract from this conclusion to note that NSF as a whole has been moving, over the past several decades, to emphasize collaborative and interdisciplinary research.

In summary, it is clear from the analysis of the set of major scientific and applied breakthroughs in atmospheric science considered in this chapter that NSF ATM has made effective use of its varied modes of support and that the balance between the modes has evolved over time in response to the needs and opportunities of the field. The committee expects that ATM will continue to evolve the balance between its modes of support as atmospheric science and its applications evolve.

3

The Changing Context for Atmospheric Science

A significant evolution and growth of the atmospheric sciences has occurred since the first National Academies' review of the status of research and education in the field (NAS/NRC, 1958). The expansion of university, private-sector, and National Center for Atmospheric Research (NCAR) research and the development of new communications and computational infrastructure, coupled with greatly expanded research and operational efforts at other agencies and in other countries, has transformed understanding of the atmosphere, created new operational observational and modeling capabilities, and changed the way in which atmospheric research is conducted. The expansion of the field has also led to significant achievements and scientific discoveries with direct societal benefits, such as decreased economic losses in a number of private sectors due to improvements in weather predictions.

New subdisciplines of atmospheric science have emerged, such as climate change and atmospheric chemistry, which grew out of an increased awareness of air pollution. The number and size of university atmospheric science programs has increased by nearly a factor of five, indicative of a more comprehensive and richer research endeavor. NCAR has grown to an institution that houses about 935 scientists and support personnel, builds and maintains observational and modeling facilities, and serves as a leader in organizing field campaigns, educational and outreach activities, and other community service efforts. Federal agencies other than the National Science Foundation (NSF), including the National Aeronautics and Space Administration (NASA), National Oceanic and Atmospheric Administration (NOAA), Department of Energy (DOE), Department of Defense

(DoD), and Environmental Protection Agency (EPA), have added to the support for atmospheric research, both internally and extramurally. These other agencies have focused efforts on their own missions and supporting research objectives (e.g., air quality) and have pioneered new approaches to research, most notably the introduction by NASA of space-based platforms for observing the atmosphere and near-space environment. International collaborations, including large multi-investigator and multinational field campaigns, now play a major role and require a significant fraction of the research budget.

In this chapter, some aspects of the evolution of the atmospheric sciences from 1958, when the National Academy of Sciences (NAS) first considered the status of research and education activities in the field, to the present are analyzed. While illustrative rather than comprehensive, this consideration of a number of key factors that influence the field—including the broader intellectual and societal context, demographics, and technology developments—has helped inform the committee's thinking about what factors are important in shaping future directions for the atmospheric sciences.

INTELLECTUAL AND SOCIETAL CONTEXT

During much of the 20th century, atmospheric scientists focused primarily on issues of weather, greatly expanding our understanding of the physical dynamics of the lower atmosphere required for weather forecasting. In the early years, the Navy, Department of Agriculture, the Army Medical Department, the Smithsonian Institution, the Signal Office, and other government programs supported research to develop accurate weather predictions for storm forecasting, aviation, and agriculture (Fleming, 1997). Indeed, the 1959 "blue book" report of the University Committee on Atmospheric Research ("UCAR," 1959) that presented the scientific rationale for the establishment of a large national atmospheric sciences research center focuses on atmospheric physics topics relevant to meteorology, balanced by a recognition of cross-disciplinary research avenues such as aeronomy, atmospheric chemistry, and the possible impact of atomic weapons detonations on the atmosphere's electrical structure. Basic and applied research in meteorology over the past several decades has contributed to remarkable advances in knowledge of the atmosphere, discoveries of relevance to scientific inquiry more broadly (e.g., the discovery of chaos theory, as described in Box 3-1), and greatly improved abilities to forecast atmospheric conditions.

Atmospheric science has been deeply rooted in practical applications since its inception, so that the need for research to meet societal expectations and to lead to progress in operations has long been an organizing principle. Indeed, it is striking that many of the topics highlighted in the

1959 NAS/National Research Council (NRC) report *Proceedings of the Scientific Information Meeting on Atmospheric Sciences* remain among the major focus areas for research and development, such as improvement of understanding and methods related to weather forecasting, pollution and its health effects, fire risk, droughts, agriculture, erosion, and water management, to name a few. Although a few topics identified in the 1959 report have so matured through technical advances that continued research is not as prominent a feature of the scientific landscape as it was in the past, these are the exceptions. Further, a number of topics have been added to the menu of societal concerns, particularly seasonal-to-interannual climate forecasting, global change, space weather, and atmospheric dispersion of chemical, nuclear, and biological contaminants.

The range of products that are needed and expected by an ever more engaged and broader public continues to expand and deepen, building upon the successes and development since the 1950s. It seems apparent that the public's interest in gaining access to the information relating to these topics has increased rapidly. Today's citizen makes greater demands on research to deliver a far larger number of user-oriented products. Examples include urban air quality forecasts, agricultural forecasts tailored to specific farming areas or crop types, as well as lightning detection systems to assist in fire risk evaluation. New warning systems, such as online access to hurricane and tornado forecasts, are also among the products that now enjoy large constituencies due to the availability of the Internet as well as the greatly improved capacity of scientists to provide increasingly accurate and ever-faster response information, enhancing public safety. These are only a few examples of the many types of products that reflect the ever-increasing pace of application of research to operations and products (NRC, 1998b).

Society also expects more finely tailored and wider ranges of information, provided in terms understandable to a broad audience. As scientific information and understanding have deepened on topics such as atmospheric pollution and climate change, there has been a far deeper appreciation of the policy relevance of atmospheric science for societal decision making (e.g., Box 3-2). Indeed, the findings of atmospheric science have provided the cornerstones for policy measures such as the Clean Air Act, the Montreal Protocol on Substances that Deplete the Ozone Layer and its subsequent amendments, and the Kyoto Protocol. Public interest in understanding how such policies work, the basis for their application, and the impact they will have has led to an increasing demand for research organizations to provide summaries aimed not just at the policymaker and other scientists, but to a far broader range of audiences, including the public, local and state governments, industry, and the education sector.

Addressing the broader impacts of research beyond advancement of knowledge has been an important thrust of NSF in recent years. All NSF

BOX 3-1
Meteorology, Chaos Theory, and Edward Lorenz

The Atmospheric Sciences have contributed many discoveries of great importance to the overall fabric of science, but perhaps none has had the impact on physics and mathematics as the introduction of chaos theory by Professor Edward Lorenz. The following discussion is taken largely from Chapter 3 of Lorenz's (1993) excellent popular book, *The Essence of Chaos*. This discussion is an interesting example of the interplay among a large international field program, large-scale numerical modeling, and the brilliant work of a single investigator.

In the early 1960s, several leading meteorologists (see Charney et al., 1966) were planning for one of the largest atmospheric field programs, the Global Atmospheric Research Program (GARP), that has been executed to date. The concept was to show how improved meteorological data would enable better weather forecasts to be made. Among the original aims of GARP was the goal of enabling two-week weather forecasts. Jule Charney was concerned, however, that the feasibility of making a two-week weather forecast might be proven impossible before the first such forecast was even attempted. During a special conference in Boulder, Colorado, Charney convinced all of the global atmospheric circulation modelers to perform experiments in which pairs of numerical forecasts were made starting from slightly different initial conditions. These experiments showed that the doubling time for small errors in the initial conditions was about 5 days.

Professor Edward Lorenz investigated the mathematical basis of such behavior using both idealized mathematical models (Lorenz, 1963) and state-of-the-art numerical weather prediction models (Lorenz, 1982). Lorenz (1963) discussed how the time evolution of a physical system may be described by a trajectory in a multidimensional phase space. He introduced an idealized set of three seemingly simple nonlinear equations in three variables as an idealization of the equations for fluid convection and showed that for suitably chosen parameters, the solutions were unpredictable in the

grant proposals are evaluated in terms of their broader impacts, which include educational objectives, broadening the participation of underrepresented groups, enhancing the infrastructure for research, wide dissemination of research results, and benefits to society. The NSF-wide small-center programs (i.e., Science and Technology Centers and Engineering Research Centers) have placed even more emphasis on education and outreach.

As the 20th century closed, the content and context of the atmospheric sciences had expanded dramatically. In addition to discussing the major issues facing meteorology, the NRC's report *The Atmospheric Sciences—Entering the Twenty-First Century* (NRC, 1998b) highlights the challenges of improving and maintaining air quality, protecting and improving ecosystems, sustaining the stratospheric ozone layer, understanding and man-sense that two solutions with arbitrarily close initial conditions soon departed from one another so that after some time there was no way to identify that the two solutions were at an earlier time very close to one another in phase space.

The implications of Lorenz's work were enormous for meteorology. It became clear what the mathematical limits for deterministic weather forecasting were, and the goal for deterministic weather forecasting was from that time on to see how much improvements in weather prediction models and observational systems could produce forecast skill for times within the theoretical limit for predictability. It also set the stage for ensemble weather forecasting which is the present state of the art in which several forecasts are run that span the uncertainties in both initial conditions and physical parameterizations so that we are not only predicting the weather but also predicting the confidence in the spread of the model predictions.

The implications of Lorenz's work were not confined to meteorology. Chaos theory found its way into mathematics and physics and other fields, and books with titles such as *From Clocks to Chaos* (Gleick and Mackey, 1988), *Chaos: Making a New Science* (Gleick, 1987), *Order Out of Chaos* (Prigogine and Stengers, 1984), and *Chance and Chaos* (Ruelle, 1991) appeared showing the generality of the concept.

Clearly, Lorenz's work had its beginnings in meteorology and had great impact on the field, but it also had far-ranging implications in areas beyond the atmospheric sciences. It was the result of a brilliant individual performing his own research, but its motivation was from a much larger community problem, whose importance was well recognized.

It should be noted that the Lorenz (1963) paper has an acknowledgement of Air Force funding for this research, but would the Air Force fund such work now? It is worth asking whether any agency other than the NSF Division of Atmospheric Science would be funding this type of research today.

aging climate variability and global change, characterizing the near space environment, and developing the ability to predict space weather. The central role of the atmospheric sciences in the addressing challenges of global environmental change was also addressed in a massive 1999 NRC report *Global Environmental Change—Research Pathways for the Next Decade* (NRC, 1999b). Both of these landmark reports emphasize the close coupling between atmospheric properties and processes occurring in the oceans, on land surfaces, in the near-space environment, and on the sun. They clearly demonstrate that atmospheric scientists need to collaborate closely with a wide range of colleagues from the physical, biological, Earth, and space sciences to meet the challenges facing atmospheric scientists during the 21st century.

BOX 3-2
Keeling's CO_2 record

Charles D. Keeling, 1928–2005
Excerpts from *Annual Review of Energy and Environment*, 1998, 23:25–82

"When I began my professional career, the pursuit of science was in a transition from a pursuit by individuals motivated by personal curiosity to a worldwide enterprise with powerful strategic and materialistic purposes. The studies of the Earth's environment that I have engaged in for over forty years, and describe in this essay, could not have been realized by the old kind of science. Associated with the new kind of science, however, was a loss of ease to pursue, unfettered, one's personal approached to scientific discover. Human society, embracing science for its tangible benefits, inevitably has grown dependent on scientific discoveries. It now seeks direct deliverable results, often on a timetable, as compensation for public sponsorship. Perhaps my experience in studying the Earth, initially with few restrictions and later with increasingly sophisticated interaction with government sponsors and various planning committees, will provide a perspective on this great transition from science being primarily an intellectual pastime of private persons to it present status as a major contributor to the quality of human life and the prosperity of nations."

"In 1953, I complete a dissertation on polymers under Dr. Dole, taking what was then the extraordinarily long period of five full years. I had acquired a working knowledge of geology, weak in laboratory and field work...."

"Although I hardly grasped it then, the opportunities for new Ph.D.s were at nearly an all-time peak. There had been a shortage of Ph.D. chemists ever since the recent world war... I was offered employment by large chemical manufacturers, most of which were located in the industrialized cities of the eastern United States... In more recent times it would have been risky to pass up such good job offers... I accepted an invitation from Professor Harrison Brown of the California Institute of Technology

RESEARCH SUPPORT

The atmospheric sciences have enjoyed a slow but steady increase in funding by NSF since the late 1950s. NSF funding for atmospheric sciences was $16.3 million (in constant 1996 dollars) in 1958, increasing to $53.9 million in 1959. The ATM budget had increased to $122 million by 1972, reaching $196 million in 2004 (Figure 3-1). Much of the budget increase that ATM has experienced since the 1980s can be traced to new funds for facilities operated by entities other than NCAR ($27 million increase since 1982) and for NSF-wide priorities, such as "Biocomplexity in the Environment" and "Information Technology Research" ($25 million increase since 1989). The core grants program and NCAR have experienced modest increases in support over the past 30 years.

in Pasadena, California, where he had recently started a new department in Geochemistry. I became his first postdoctoral fellow."

"With Professor Brown's consent, I postponed the study of uranium in granite and set about building a device to equilibrate water with a closed air supply. I acquired a hand-operated piston pump. Through a nozzle it could spray water from a natural source onto the wall of the glass chamber to bring about a thermodynamic equilibrium between the carbon dioxide dissolved in the stream of water and gaseous CO_2 in the chamber.... I did not anticipate that the procedures established in this first experiment would be the basis for much of the research that I would pursue over the next forty-odd years."

"The highly variable literature values for CO_2 in the free atmosphere were evidently not correct. Rather a concentration of 310 ppm of CO_2 appeared to prevail over large regions of the northern hemisphere. I had detected this near-constancy under the implausible circumstances of studying air in old-growth forests where variability was to be expected. By 1956 my broader findings of surprising near-constancy seemed to me secure enough to communicate them to others...."

"The consumption of fossil fuel has increased globally nearly three-fold since I began measuring CO_2 and almost six-fold over my lifetime."

These early studies of atmospheric CO_2 concentrations by Charles D. Keeling in California provided a foundation for the establishment of the now famous Mauna Loa CO_2 measurements. The autobiographical text from Dr. Keeling is provided as an example of how atmospheric science has evolved, and underscores the importance of working across scientific disciplines and investing in transformative and sometimes risky work for individual investigators.

This funding is currently directed to the modes of support including core grants, university facilities, NCAR facilities and science, and NSF priorities, as shown in Figure 1-2. These modes overlap in many ways, for example, because facilities are integral to the research process. Over these 30 years, core research has decreased from 50 to 38 percent of the overall ATM budget, and support for science at NCAR has decreased from 23 to 18 percent of the ATM budget. However, given the overall increase in the ATM budget, NSF core grant support has remained about constant in total dollars. At the same time, support for facilities at NCAR and at universities has increased from 23 to 33 percent of the ATM budget. Thus, facilities support has increased faster than core grant support, most likely due to the increasing sophistication of computing and observing capabilities. The com-

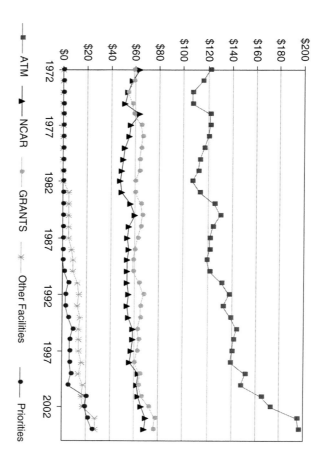

FIGURE 3-1 ATM funding for the atmospheric sciences since FY 1972 in constant 1996 dollars. The NCAR numbers include support for both science and facilities housed at the center. "Other facilities" refers to support for facilities operated by institutions other than NCAR. "Priorities" refers to NSF-wide initiatives, such as "Biocomplexity in the Environment" and "Information Technology Research."

mittee notes that the availability of facilities creates research opportunities for individual investigators.

Many of the advances in the atmospheric sciences have been enabled by the availability of sophisticated, and expensive, facilities. These include supercomputers, research aircraft, and high-power radar systems, so it is not surprising that during the last 30 years, the fraction of ATM funding devoted to facilities has grown from 23 to 33 percent of its budget. Very valid arguments can, and will, be put forth for ATM purchasing bigger, and more expensive, computers and very valuable, and expensive, observing facilities in the coming years. At the same time, if ATM's budget is not increasing faster than inflation, the funds to purchase and maintain these facilities will have to come from the research budget. Thus, ATM will be presented with a dilemma of how to make the trade-off between investments in "tools" at the expense of funding people when both will be necessary to generate and implement ideas. Ideally these difficult decisions

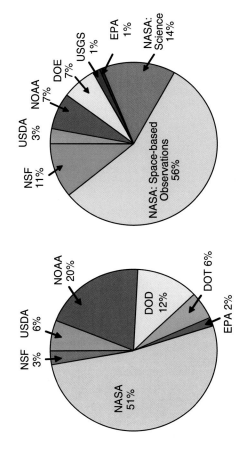

FIGURE 3-2 (Left) FY 2004 funding for weather and space weather research by the 10 agencies surveyed by the OFCM. The overall funding by these agencies for this year totals about $503 million. Note that the NSF funding only includes the foundation's contributions to space weather research and the U.S. Weather Research Program, which together total about $14 million. The NASA proportion of the OFCM funding is composed of the estimated meteorology share of the supporting research and analysis programs as well as Earth Observing System (EOS) and Earth Probe instruments, EOS science, and the EOS Data Information System elements of the NASA Office of Earth Science budget (OFCM, 2004). (Right) Estimated FY 2003 budget for atmospheric-related climate change research (i.e., the atmospheric composition, climate variability and change, carbon cycle, and water cycle program areas) by the 13 agencies of the CCSP (CCSP and SGCR, 2004). Total funding for these program areas is approximately $1.4 billion.

will be made in the framework of ATM's strategic planning and with input from the broad atmospheric sciences community.

Other agencies have experienced much larger fluctuations in their extramural funding for atmospheric science. It is not easy to track down exactly how much each agency spends on atmospheric research; Figure 3-2 shows efforts by the Office of the Federal Coordinator for Meteorology (OFCM) and the U.S. Climate Change Science Program (CCSP) to sum up the contributions of different agencies to research relevant to their individual mandates. Note that the agencies also support research on air quality and solar sciences, which neither of the charts in Figure 3-2 includes. These budgets include both intramural and extramural support for research. ATM is a relatively small player overall, but plays a significant role in supporting university and other extramural research. In the last 5–10 years, NASA and

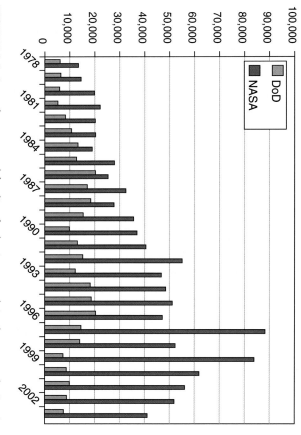

FIGURE 3-3 Annual extramural funding for basic research in the atmospheric sciences at DoD and NASA (NSF, 2004).

DoD have decreased their support for basic research in the atmospheric sciences (Figure 3-3). This reduced funding for basic atmospheric research by other federal agencies will likely cause more and more of the community to turn to ATM for basic research funding.

The committee believes that this evolution of atmospheric science research since 1958 introduces not only new opportunities but also new challenges. For example, five years of steady growth in NSF budgets have given way to a new period of limited budget growth, while support of atmospheric science research by other federal agencies exhibits considerable volatility. The constrained budget environment combined with the expanded scope of scientific questions have increased the need for interagency and international coordination.

DEMOGRAPHICS

NAS (NAS/NRC, 1958) concluded that there was a strong need for more professionals in the atmospheric sciences. At the time, only about 10 to 15 doctorates were awarded each year. By the late 1970s, an average of 84 doctorates a year was awarded by a greatly expanded number of university atmospheric sciences departments in the United States, meeting the needs for professionals in the field at that time (*http://www.ametsoc. org/EXEC/TenYear/figs.html*). Table 3-1 provides a number of indices for the growth in the atmospheric sciences research community. The table illustrates the significant expansion of educational efforts, professionals, and research funding over the past four decades, although it is difficult to pin down the exact size of the community because of its diversity.

The size of the research workforce in atmospheric and related sciences seems to have been leveling off since the 1990s because of lower interest in the physical sciences, the growth of research programs overseas, and movement of some of the Ph.D. population to the private sector (Hoffer et al., 2001; Vali et al., 2002). On average, 133 atmospheric science doctorates were awarded annually in the late 1990s. The number of applicants to atmospheric science graduate programs declined between 1995–1996 and 1999–2000 (Vali et al., 2002), but increased slightly as of the 2002–2003 academic year, the most recent year with available data (Vali and Anthes, 2003). In the coming years, there is a projected shrinkage of the science and engineering research labor pool through retirements (NSB, 2002) coupled with a projected growth in science and engineering career opportunities. It is not clear exactly how these broad trends for the physical sciences and engineering might impact the atmospheric sciences.

Note that not all atmospheric scientists are trained in atmospheric science, meteorology, astronomy, or Earth science departments. In particular, atmospheric chemists and cloud/aerosol microphysicists may be enrolled in chemistry, physics, applied science, chemical engineering, aerospace/ mechanical engineering, civil/environmental engineering, or public health programs. Solar physicists, aeronomers, and other near-space scientists may be trained in astronomy, physics, chemistry, or electrical engineering departments. Those who study marine meteorology or interactions between the atmosphere and the ocean may enroll in marine science departments. ATM supports research in all of these academic enclaves.

Along with efforts to increase the size of the atmospheric sciences workforce, the meteorological community has worked to make the production and communication of weather information more professional (NRC, 2003b). Private-sector meteorology began in earnest in this country shortly after the end of World War II, when several thousand meteorologists trained to support the massive aviation activities of the U.S. armed forces left government service eager to apply their newly acquired skills (Mazuzan, 1988). The Weather Bureau made the decision to permit its weather data to be used by the emerging private sector, and the first group of private meteorological companies began operating in 1946. The emerging television industry was a natural outlet for weather information and forecasts, and the decision by the Weather Bureau that government employees would not provide television weathercasts prompted the development of an influ-

TABLE 3-1 Overview of Trends in Demographics and Research Support in the Atmospheric Sciences

Year	Number of UCAR Member Institutions[a]	Number of Atmospheric Science Ph.D.s Granted per Year[b]	Number of AGU Meteorology Section Members[c]	Number of AGU Atmospheric Science Section Members[c]	Number of AGU Space Physics & Aeronomy Section Members[c]	Number of AMS Members[d]	Total Annual NSF Support for Atmospheric Science (millions of 1996 dollars)
Late 1950s	14	10	1,700 (in 1958)	—	—	7,000	16.3 (in 1958) 53.9 (in 1959)
1976–1980	46	84	—	1,600[e]	1,610[e]	9,000	119
1996–2000	68	133	—	5,300	3,430	12,000	144

[a]Data from NAS/NRC (1958), University Corporation for Atmospheric Research (UCAR) archives for 1976–1980, and UCAR Web site for 1996–2000.

[b]Data from the NSF Science Resource statistics.

[c]Data provided by American Geophysical Union (AGU). In the late 1950s, atmospheric science, space physics, and aeronomy were all grouped into a meteorology section. Note that approximately 30 percent of AGU membership in recent years is from outside the United States.

[d]Data from *http://www.ametsoc.org/EXEC/TenYear/figs.html*; the current membership of the American Meteorological Society is distributed almost equally among the private, public, and academic sectors.

[e]Averages are for 1978–1981.

ential component of the private sector—broadcast meteorology—as well as competition among weather information providers to develop better visualizations and other products for the weathercasters. The American Meteorological Society (AMS) started its Board on Broadcast Meteorology in 1957 to encourage more science-based programming (AMS, 2006). Today, there are over 250 private meteorological companies in this country providing operational forecasts, consulting services, data services, and research and development.

In addition, much of the growth in the graduate-level workforce over the past 50 years has involved (directly or indirectly) federally supported jobs at the research universities, NCAR, and the mission agencies (principally NOAA, NASA, the EPA, and DOE). For the most part, the level of NSF support devoted to graduate-level training has been appropriate. So far, the field has experienced no acute shortages of well-trained professionals, nor has it experienced periods of oversupply of graduates. There are concerns that the increasing pressures on the federal budget may reduce the number of federally supported jobs in the field, leading to an oversupply over the next few years. However, more graduates are working for the private sector: AMS statistics show 52—or slightly more than 25 percent of the Ph.D.s from 1997 through 1999—chose a career in the private sector, more than those that chose university faculty (32) to civilian government (46) positions. Many of the private-sector jobs are not research jobs. Indeed, anecdotal evidence indicates that some of these Ph.D.s pursued opportunities outside the field, using their research training and skills.

The community of atmospheric scientists in the United States has long included significant participation by individuals from other nations. In fact, 30 to 40 percent of Ph.D. degrees over the past decade have been awarded to students with temporary visas (WebCASPAR, 2006). In recent decades, students have come to the United States to train; the number of foreign-born graduate students in physical sciences and engineering has increased both in absolute numbers and as a percentage. Growth continued into the mid 1990s, when it reversed (Hoffer et al., 2001). The downturn is related to the increase in opportunities for university training abroad (NSB, 2003) and, since 2001, there have been modest impacts on graduate school enrollments from increased restrictions for foreign students traveling to the United States (NRC, 2005b). This downturn in applications from abroad combined with the increasing difficulty in recruiting American students to the sciences seems to jeopardize the U.S. standing as a leader in science and innovations (NRC, 2005b).

As in most scientific fields, the number of women and minorities in the atmospheric sciences has increased over the past decades, although there is room for further improvement. Since the beginning of the 21st century, an equal number of men and women have earned Ph.D.s; however, women are

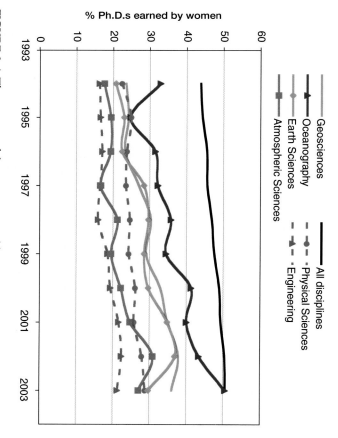

FIGURE 3-4 The percent of doctorates earned by women in Atmospheric Sciences, Earth Sciences, Oceanography, Geosciences, Engineering, Physical Sciences, and all university disciplines (Data Source: WebCASPAR, 2006).

still underrepresented in most physical sciences (Figure 3-4). In 2003, 28 percent of Ph.D.s in the atmospheric sciences were awarded to women, an increase from about 20 percent in the mid 1990s. Minority representation in the atmospheric sciences remains low (Table 3-2). Only 1.3 percent of Ph.D.s in the atmospheric sciences were awarded to African Americans and 1.4 percent to Hispanics from 1993 to 2004. Atmospheric science attracts fewer minorities than physics and the other Earth sciences with which the field competes in recruiting applicants from the existing science student pool, perhaps because students are more likely to be exposed to physics, chemistry, and biology in their high school education (e.g., Barstow and Geary, 2002). This recruitment challenge is complicated by the lack of exposure to atmospheric sciences by talented undergraduates at minority-serving institutions, which may often offer no more than an introductory survey course in the field.

The urgency of recruiting minorities will be amplified as the ethnic makeup of our nation changes. The percentage of non-Hispanic American whites, who make up the bulk of the scientist population, is projected to

TABLE 3-2 Ten-Year Average (1994–2003) of Ph.D.s Granted to Minorities in Engineering, Physical Sciences, Atmospheric Sciences, Geosciences, and Social & Natural Sciences Combined, and Minority Categories as a Percentage of Total U.S. Population

	Engineering	Physical Sciences	Atmospheric Sciences	Geo-sciences	All Sciences	% US Population
African American	2.9	2.5	1.3	1.5	3.8	12.8
Native American	0.4	0.3	0.1	0.4	0.5	1.0
Asian & Pacific Islander	21.2	14.7	15.2	9.8	12.7	4.4[a]
Hispanic	3.3	3.0	1.4	3.0	3.8	14.1
Caucasian	69.4	76.8	79.4	82.5	76.7	67.4[b]
Other	2.8	2.7	2.6	2.8	2.6	1.5[c]

[a]Total of Census Bureau categories "Asian" (4.2%) and "Native Hawaiian and Other Pacific Islander" (0.2%).

[b]Census Bureau category "White Persons, Not Hispanic" ("White Persons," 80.4%).

[c]Census Bureau category "Persons Reporting Two or More Races."

SOURCES: WebCASPAR (2006); *http://quickfacts.census.gov/qfd/states/00000.html*.

drop from 77 to 53 percent between 1999 and 2050. This is slightly offset by a projected increase in Asian Americans, who participate significantly in science, from 4 to 9 percent. However, the percentage of underrepresented groups in science is expected to increase in the general population, largely because the percentage of Hispanics in the United States is projected to double, from 12 to 24 percent. The percentage change of African Americans and Native Americans is projected to be small (12 to 13 percent and 0.7 to 0.8 percent, respectively) (NSF, 2003).

The number of minority Ph.D.s in the atmospheric sciences, as well as other physical sciences, has increased over the past 10 years. Atmospheric sciences could increase its efforts in competing more effectively against physics and the other Earth sciences in recruiting students from the existing science student pool, using models such as UCAR's SOARS® program. Nevertheless, it would not resolve the overall difficulties of recruiting minorities to graduate programs in sciences. To make any real and lasting progress, a strengthening of the K-12 mathematics and sciences curriculum, particularly in schools with high numbers of minority students, would be necessary (NRC, 2005b). As mentioned previously, an increasing deficit in the available science talent pool seems to arise, suggesting an even greater competition for minority students in the future (NRC, 2005b).

OBSERVATIONS: TECHNOLOGY DEVELOPMENT AND EMERGENCE OF FIELD PROGRAMS

Atmospheric research, operations, and products rely heavily on observations of the state and composition of the atmosphere, oceans, and land surfaces. Evolution of our understanding and forecast capabilities have been associated in part with new measurement capabilities resulting from new sensors, new observing platforms, and systems of instruments within networks. Automation for remote observations, reduction in size of instrumentation, computational processing, easy access to data and information, new signal processing capabilities for analysis, and improved visualization technology have all provided us with the tools to produce science products for research, operations, and user information services.

Major advances in technology since the 1950s include satellite observing platforms and instrumentation; new Doppler radars for the lower and upper atmosphere; and the ability to measure processes, not just state variables. Satellites have led to great improvements in the study of evolving weather patterns and the distribution of atmospheric pollutants, especially in data-sparse regions. The development and implementation of satellite-based observing platforms has largely been the purview of NASA and NOAA, while ATM has been the primary funding source for non-space-based platform instrumentation development. A significant portion of the NSF-supported instrument development has taken place at NCAR, where a major, centralized national facility was formed. This facility consists of unique observing systems and platforms otherwise not readily accessible to NSF-sponsored Principal Investigators because they would be difficult for any single university person or group to develop. One such example of this type of platform is the High-Performance Instrument Airborne Platform for Environmental Research (HIAPER), which debuted in 2006 as the nation's most advanced environmental research aircraft. This aircraft and many other observing systems are maintained by NCAR and supported for field programs by the NSF deployment pool.

Major technical achievements in incoherent scatter radars along with the siting of these radars in a longitudinal network have enhanced process understanding of geomagnetic storms, Sun-Earth connections, and ionospheric disturbances. Combined with models, these technical advances have provided the framework for space weather forecasting. A variety of smaller upper-atmosphere radars have emerged, providing sometimes the only observational information on the dynamics of the electrically neutral mesosphere-lower thermosphere, leading to major revolutionary thinking about the theory of circulations in the upper atmosphere. Associated optical instrumentation development, especially resonance and Rayleigh-scatter lidars, has led to new measurements of chemical constituents in the upper atmosphere.

The development of compact, robust, highly sensitive real-time trace-species and fine-particle sensors, many based on spectroscopic or mass spectrometric measurements, has allowed the deployment of multisensor suites on mobile platforms (aircraft, balloons, ships, vans) capable of mapping ambient atmospheric pollutant concentrations and characterizing surface sources and sinks (Kolb, 2003). The development of fast trace-gas and fine-particle sensors has also enabled the direct measurement of surface emission and deposition fluxes, using micrometeorological techniques from flux towers and low-altitude aircraft.

Typical atmospheric observational studies have involved a mix of routinely available measurements and those collected as part of a field program. Since the 1950s, the level and sophistication of routinely available observations has expanded. The U.S. Weather Service modernization provided improved radar coverage starting in the 1990s. Longer-term field campaigns, such as the DOE Atmospheric Radiation Testbeds, have provided continuous streams of measurements in the central United States, the Pacific, and Alaska. The Tropical Ocean-Global Atmosphere Tropical Atmospheric Ocean (TOGA-TAO) array provides surface atmospheric and oceanic data from the tropical Pacific. Starting in the 1990s, the U.S. commercial aircraft fleet started sampling temperature and wind, and humidity measurements are now being taken. Satellites supply a rich mix of data that characterize the surface, ocean currents, atmosphere, thermal stratification of the atmosphere, cloud cover, tropical precipitation, aerosol distribution, and trace gas concentrations. Assimilated into numerical models, these data can provide a reasonably good picture of the systems that provide our day-to-day weather and motions of longer time and spatial scales. Providing a framework for analyzing historical data are up to four decades worth of dynamically consistent data produced by reanalysis efforts.

In recent years, the importance of climate change in the atmospheric sciences has created new observational demands for monitoring of the atmosphere, in particular, for sustained observations with global coverage. Satellite-based observations have provided major advances, but suffer from lack of continuity and related problems of calibration among instruments, necessitating continued investment in, and use of, in situ platforms. The need for enhanced monitoring overall requires continued attention to the development of instruments that are more robust, numerous, lightweight, easily deployable and maintained, and less expensive. NASA and NOAA are major players in the monitoring arena, but further work in this area is needed to ensure an adequate future climate observing system (NRC, 1998b, 1999a).

Although operational and monitoring data are often sufficient to study larger-scale motions, field programs are needed for coordinating additional measurements to address specific questions regarding atmospheric processes not resolved by models, and requiring measurements not routinely made (e.g., Box 3-3). Making instruments and platforms available to the community to collect these measurements was a major reason for the establishment of NCAR. Many important field campaigns over the past 45 years have been relatively small, involving fewer than a dozen investigators and focusing on short-term atmospheric processes over a relatively limited geographical area.

In 1974, ATM and NCAR were major players in the GARP Atlantic Tropical Experiment (GATE)—the first large, international field program—by providing three aircraft and significant support in planning and logistics. Since GATE, the number of large and multinational field programs addressing tropospheric research questions has multiplied. Many current lower-atmosphere field programs address a broader spectrum of disciplines (e.g., oceanography, soils, ecology, hydrology, and chemistry), and there is pressure to extend to longer timescales, largely in response to increased focus on climate issues, biogeochemical cycles, and the water cycle. More frequent field campaigns and larger field programs now compete for resources. Increasingly, investigators propose research activities that are synergistic with the initial phase of a field campaign, thereby increasing its logistical complexity and scientific scope. Furthermore, non-U.S. scientists are assuming more leadership roles in large, international field campaigns. Examples include the African Monsoon Multiscale Analysis Experiment, which is sponsored by the European Union (EU) and led by scientists in France, and the Atmospheric Brown Clouds project, sponsored by NOAA and the United Nations Environment Programme and led by German and U.S. scientists.

The upper-atmospheric research community also conducts field campaigns, often planned around fixed observing facilities, such as is the case for the Maui Mesosphere and Lower Thermosphere campaign. There have also been a series of field campaigns over the past two decades that have been supported by both monthly "World Days," when upper atmosphere observations are coordinated, and there are longer periods of continuous observations by the network of incoherent scatter radars.

Numerical modeling has played an increasingly important role in developing observational strategies and subsequent data analysis. Starting in the 1970s, numerical modelers influenced the location, type, and frequency of observations, the design of field programs to test parameterization schemes for moist convection and the forecasts used for measurement strategy. Now, the roles of models and observations are intimately entwined. Aircraft and other observational platforms may be deployed to fill in a data void in an area that numerical simulations show to be a source of forecast uncertainty or where convective storms are likely to originate. Finally, detailed datasets are assimilated into models to provide a more complete picture of the phenomenon being studied.

ATMOSPHERIC SCIENCE OBSERVATIONAL TOOLS AND TRAINING

Despite tremendous advances in the sophistication and accuracy of computational models designed to describe and predict atmospheric processes, progress in the atmospheric sciences is highly dependent on the quality of instruments and observational systems available to measure the physical and chemical properties of the atmosphere. Progress will also depend on the instrumental and observational skills of new generations of atmospheric scientists who need to master ever more sophisticated arrays of in situ and remote atmospheric sensing tools.

Effectively nurturing the development of new measurement methods and tools and ensuring the measurement skills of young atmospheric scientists are complex tasks that present special challenges to the atmospheric sciences community and the agencies that support atmospheric research and education. Some of these challenges and recommendations for meeting them are detailed below.

Instrument and Measurement System Innovation

Several previous NRC reports have noted the importance of innovative measurement tools for atmospheric science and the challenges associated with their development. For instance, in a 1990 status report on the Global Tropospheric Chemistry program the NRC's Committee on Atmospheric Chemistry noted "the inability to measure many critical trace species and their transport fluxes with sufficient reliability, accuracy, and temporal or spatial resolution" (NRC, 1990). That report went on to observe that too few research groups are mounting advanced development efforts for measuring species important in atmospheric chemistry because "advanced instrumentation development is an extremely challenging and risky activity."

The challenges to successful instrument development listed in that report included: (1) the need for a development team with diverse talents, including mastery of specific scientific measurement techniques and comprehensive engineering design and implementation skills spanning topics such as electro-optics, electronic and computer control, fluid mechanics,

BOX 3-3
A History of Field Experiments

Joachim Kuettner, Chair
Atmospheric Sciences and International Research, UCAR/NCAR
Ph.D., Physics (Meteorology), University of Hamburg, Germany

My career in science has spanned much of the last century. My passion has been to combine both science and flying. Beginning with a study of the newly discovered lee wave in Germany's Riesengebirge Mountains, I obtained a doctorate in physics from the University of Hamburg in 1939. I have worked with numerous aircraft for research, including low- and high-altitude, motorized and gliding aircraft. In pursuing this love of flight, I established several world and national gliding records, including an absolute altitude gliding record for high altitude (43,000 ft [13 km]; still the German gliding high-altitude record).

Before becoming involved in many field projects associated with NSF, I joined Wernher Von Braun's team at NASA's Marshall Space Flight Center in Huntsville, Alabama (1958-1965), and became the Center's Director of the flights of the U.S. astronauts. Subsequently, I was Director of the Apollo Systems Office, responsible for the integration of the Apollo spacecraft and the Saturn-V rocket for the lunar landing. After I left Huntsville I became the Chief Scientist at the National Weather Satellite Center in Washington; and in 1967, Director of Advanced Research Projects at NOAA in Boulder, Colorado.

The many atmospheric researchers who have flown on research aircraft have learned that you don't really know the Earth's atmosphere until you have experienced it personally in flight. For example, during the "Mercury-Redstone" Project (the first space flights of the U.S. astronauts, Alan Sheppard and Virgil "Gus" Grissom). Subsequently, the Electra penetrated the monsoon front over the Arabian Sea about two days prior to landfall in India. Our colleagues from India who had devoted a lifetime to monsoon studies were almost delirious at seeing the inner structure of the phenomenon of their life's work. In addition to facility support by NSF, whenever I needed a break as scientific director, I would invite our NSF program managers, Jay Fein (summer MONEX in India) and Dick Greenfield (winter MONEX in Malaysia), to take over the project management and the planning meetings. They did a remarkable job in the field.

Prior to MONEX, in the early 1970s, all member countries of the World Meteorological Organization (WMO) agreed to implement GARP, the Global Atmospheric Research Program. Its first field project, the GARP Atlantic Tropical Experiment (GATE) in 1974, was a huge undertaking involving almost 4,000 participants from 70 nations. The USSR participants were particularly thrilled to engage in the active, open exchange of ideas that characterized GATE. The NSF's participation in the GATE experiment was very important to its success.

Among the many worries that I had as the WMO-appointed head of GATE was the daily deployment of the flexible observing systems, such as 13 aircraft and 39 ships, for a host of competing scientific objectives. Should the decisions be made in a more military fashion, by a "czar," or more democratically, by a mission-planning team? Could such a team act in the short time available—usually about one to two hours? Through simulations of mission planning for various, sometimes surprising, scenarios, conducted at NCAR and attended by the lead scientists from several nations (United States, U.S.S.R., United Kingdom, France, and Germany), we created a congenial and efficient "Mission Selection Team" that set the standard for practically every international experiment since. Following GATE, I planned and directed for WMO the aforementioned Monsoon Experiment. That was followed by ALPEX, the Alpine Experiment, which explored the airflow over and around mountain ranges.

Since 1985, my home base has been the National Center for Atmospheric Research (NCAR) and the University Corporation for Atmospheric Research (UCAR) in Boulder, Colorado, where I have been associated with many major field projects, such as GALE (Genesis of Atlantic Lows Experiment, 1986), TAMEX (Taiwan Area Mesoscale EXperiment, 1988), THERMEX (Thermal Wave Experiment, 1989), TOGA-COARE (Coupled Ocean–Atmosphere Response Experiment, Australia, 1992), CEPEX (Central Equatorial Pacific EXperiment, 1993), INDOEX (Indian Ocean [aerosol] Experiment, 1999), MAP (Mesoscale Alpine Program, 1999), and T-Rex (Terrain-induced Rotor Experiment, 2005–2006). In 1994, NSF created a Distinguished Chair for Atmospheric Sciences and International Research at UCAR/NCAR, which remains my current position.

The Central Equatorial Pacific Experiment (CEPEX), mentioned above, was a good example of the close cooperation in the field between the NSF Program Director and the scientific community. CEPEX focused on surface temperature regulation in the western Pacific, and was led by Veerabhadran Ramanathan. Since Ramanathan had never directed a field project, Jay Fein suggested that I help him lead the experiment. Ramanathan and I were quickly able to communicate on the same wavelength and lead the project together. This collaboration among Ramanathan, Jay Fein, and myself has continued to this day, and has led to the development of unmanned aerial vehicles to study the role of aerosols in cloud formation and climate over the ocean.

Looking back on a long professional life, it appears that I started my atmospheric research on mountain waves and rotors, and have just completed a research project on the same subject, the T-Rex project. In the 1930s, I was puzzled by observations of rotors and hope to have found some answers through the 2006 T-Rex project. It should be mentioned that T-Rex was the first operational project for the new NCAR aircraft, HIAPER, recently acquired through NSF's efforts.

structural mechanics, and thermal control; (2) the necessity to minimize size, weight, power consumption, consumable gases and chemicals, and cryogens; (3) the requirement of reliable field calibration. Further, it was noted that development time scales to move an innovative technique from laboratory proof-of-principle to a reliable field measurement tool was typically seven to ten years, a time scale difficult to manage with typical two- to three-year research grants, normal graduate student and junior faculty time frames, or private company time-to-market constraints. Finally, the 1990 report went on the recommend that agencies supporting atmospheric chemistry research to "encourage good, innovative instrument development proposals" and "that these projects can be viewed as a key R&D portion of an atmospheric research program and should be a significant (10 to 15 percent) of each agency's overall budget." It also recommended that federally funded laboratories with ongoing instrument development programs be encouraged to form partnerships with university and private-sector laboratories, noting that such arrangements might encourage students to take on instrument development projects because they would be collaborating with successful instrument-oriented professionals (NRC, 1990).

The need for innovative observational tools was also highlighted in the NRC Board on Atmospheric Sciences and Climate's report *The Atmospheric Sciences—Entering the Twenty-First Century*, which listed the development of new observational capabilities as an "Atmospheric Science Imperative." That report stated, "the federal agencies involved in atmospheric science should commit to a strategy, priorities, and a program for developing new capabilities for observing critical variables, including water in all its phases, wind, aerosols and chemical composition, and variables related to the phenomena in near-Earth space, all on spatial and temporal scales relevant to forecasts and applications" (NRC, 1998b).

In addition, a 1999 report prepared under the direction of the NRC's Climate Research Committee focused on needed upgrades in the climate observing system (NRC, 1999a). This report called for agencies involved in the U.S. Global Climate Research Program (USGCRP) to "establish a funded activity for the development, implementation, and operation of climate specific observational programs" as a way of "providing essential additional capability to operational observing systems." This activity would include the identification and measurement of "critical variables that are either inadequately or not measured at all."

NSF/ATM has played a leading role in funding global atmospheric chemistry, meteorology, aeronomy, and climate change research; including a key role in the USGCRP. While other agencies, such as NASA, NOAA, and various DoD agencies are assigned the lead role in developing satellite remote sensing systems, NSF/ATM has played a strong role in developing in situ as well as ground and airborne remote sensing observational tools. However, the main NSF funding paradigm of grants to individual academic investigators is often not consistent with the wide skill sets and long time scales required for successful observational tool development and deployment.

Given the ongoing need for innovation in observational instruments and systems, detailed above, additional effort may be required. The committee notes that traditional, non-ATM-specific NSF instrumentation activities can be useful. For instance the Major Research Instrumentation program now has an instrument development component, and, as noted in our interim report (NRC, 2005e), the NSF Small Business Innovation Research and Small Business Technology Transfer Research programs have produced valuable atmospheric science instruments. Instrument development partnerships among interested university groups, private sector organizations, and large government and federally funded research and development center laboratories could also be an effective way to access the full range of scientific and engineering skills and experience with field measurement requirements necessary for successful instrument and measurement systems development.

Training in Observational Tool Development and Utilization

A key to continued success in the earth sciences is the continued access to high-quality observations, use of multiple observation datasets, and the ongoing development of new tools. The next generation of weather radar technology, the development of new space-based instruments, and the growing sophistication associated with data processing, visualization, and analysis requires educating the next generation of scientists and engineers who are equipped with the knowledge to integrate earth science problems with engineering observational solutions. The nature of instrument development and prototyping is changing and that requires more partnership and more involvement with the private sector.

It has been shown that universities increasingly are not investing in education programs in the observational aspects of the science for various reasons (Serafin et al., 1991; Takle, 2000). The Takle article provides a concrete set of recommendations for enhancing university instruction in observational techniques. A summary of potential actions includes:

• Provide opportunities for faculty and students to participate in NCAR and other instrumentation development programs and encourage active engagement in such programs

• Consolidate and leverage the COMET and Unidata resources as well as other instrumentation and observational educational materials for classroom use at a broad array of higher education institutions

- Seek opportunities to collaborate with other universities and with other geoscience and environmental science programs and departments in the development and implementation of instrumentation courses

- Prepare videos and electronic media on specialized instrumentation and platforms for use in college and university education programs

- Develop Web-based materials or supplements that can be shared and used at other colleges and universities

- Provide field program fellowships and opportunities for students to obtain hands-on experience with instrumentation

While there are opportunities for undergraduate and graduate students to participate in NSF-funded research projects that utilize or develop observational tools, it is rarer to have courses that provide the concepts of engineering design or data processing. The challenges of providing such education and training at a single university noted in the above articles may be overcome if a more community approach is used in the development of course materials that serve as modules and use of information technology for the delivery of courses. This approach provides for the opportunity to leverage existing materials not only in universities but also those developed at the national center (COMET and Unidata) and through the AMS. In addition several workshops or "schools" that present engineering and observational tool fundamentals have been developed and implemented (e.g., NCAR workshops, International Radar School) but have not been propagated into more formal coursework in our university curricula. The development of good online material that can be shared nationally as well as select fieldwork sites that encourage hands-on engineering internships for students should be the topic of an NSF-sponsored collation effort between atmospheric science and engineering departments, the national center, the AMS, and other federal laboratories that engage in observational tool development.

INFORMATION TECHNOLOGY AND COMPUTATIONAL MODELING

The extraordinary evolution in information technology over the last 50 years has had a huge impact on the atmospheric sciences. Roughly speaking, computational capability has advanced at nearly a hundred-fold per decade throughout the entire time period. Associated with this are increases nearly as great in internal memory, data storage, and data transfer. The advent of the Internet has served to connect the community in unprecedented ways, and presently allows practical exchange of vast amounts of information. These changes have allowed an entirely new dimension of research—that of simulation and prediction—to join theory, observation, and analysis in underpinning the science. Numerical weather, climate, and air pollution "experiments" may now be conducted in an environment that can be controlled in ways not available naturally, and in great numbers compared with what can be observed in nature. Computational models allow a new means to learn as well as a new means to harness existing knowledge, toward the development of the best possible operational and research products. For example, efforts are now under way to improve seasonal climate predictions and make them relevant to numerous real-world applications (e.g., Box 3-4).

The field of data assimilation has emerged and is just starting to fulfill its promise of improving prediction (see also Box 3-4). By the 1980s, the models were good enough to "accept" specially targeted observations (e.g., see Case Study 2 in Chapter 2) to improve hurricane forecasting. Modern data assimilation lies at the intersection of analysis and simulation, and is a critical part of both research and operational prediction. It is one of the most demanding and resource-intensive aspects of modern weather prediction. Beginning with the Fronts and Atlantic STorms EXperiment (FASTEX) in 1997, several field programs have investigated the impact on model forecasts of observations focused on locations found by adjoint models to produce the most forecast error. In the near future, enhancing use of satellite data in model-specified geographic areas to increase the certainty of forecasts will become even more promising because of easier availability and faster response time. The concept of climate "reanalyses," that is, analyses of past observations using current models and assimilation methodologies optimized to represent climate parameters, is relatively recent but has provided extremely important products for research, despite known difficulties.

The use of computer simulations as a tool to understand the Sun and the space environment has grown markedly in the last two decades. Models have been developed to study the aspects of the solar interior and to predict the intensity of the sunspot cycle (see Case Study 9 in Chapter 2). Simulations of the Earth's magnetosphere and its interaction with the solar wind are now able to reproduce real events and, in the future, will be able to provide predictions of space environmental conditions. The NSF has funded a Science and Technology Center, the Center for Integrated Space-Weather Modeling that is developing a set of coupled codes extending from the surface of the Sun to the upper atmosphere of Earth. Techniques developed in tropospheric weather modeling, particularly data assimilation, are increasingly used in space physics. For example, the Space Environment Center assimilates the total electron content over the United States in near-real time to make predictions, and the solar-cycle model of Dikpati et al. (2006) uses sunspot-intensity data from the previous three solar cycles. ATM has also supported community access to space physics models by providing partial

BOX 3-4
Atmospheric Reanalyses and Dynamical Seasonal Climate Prediction

Jagadish Shukla, Professor
George Mason University
Ph.D., Banaras Hindu University, Geophysics

The NSF is unique among the federal agencies that fund academic research because it has the flexibility to entertain and support highly innovative basic research ideas that might be considered high-risk in the mission-driven agencies. I can think of no better example of that approach than the Atmospheric Sciences Division, with which I have worked for more than 20 years. In my own case, there have been several occasions when I have brought ideas to the NSF and always received a respectful hearing and an indication of interest. The following two instances are examples of many success stories that have been the result of the flexibility and vision of NSF and its program officers.

Reanalysis

In the mid 1980s, I became convinced that it was possible and essential to produce a reanalysis of data representing the four-dimensional nature of the global atmosphere for the past half-century. The technology available at that time was adequate to analyze observations of the global atmosphere using a state-of-the-art data assimilation system such as was in use for operational Numerical Weather Prediction (NWP) at the European Centre for Medium-Range Weather Forecasts (ECMWF) or the U.S. National Meteorological Center (now the National Centers for Environmental Prediction, NCEP). I was also convinced that such assimilation would be far superior to the analyses made in real time over the period since NWP was initiated in the 1950s. The data archives were adequate to uniquely define the dynamic and thermodynamic state of the near-surface and upper atmosphere every day for the period since rawindsondes were in routine use, which began about the time of the end of World War II. I started a campaign to persuade the centers involved in NWP around the world to consider undertaking such a task.

After several unsuccessful attempts to get the backing of the operational agencies, I approached the Atmospheric Sciences Division of NSF (Jay Fein) with the idea of conducting a pilot project as a proof-of-concept for reanalysis. The reviewers of the proposal were impressed and we received funding to produce a multiyear reanalysis of observations that proved to be demonstrably superior to the analyses available from the real-time NWP archive. The results eventually appeared in a 1994 article. Consequently, NCEP, in partnership with NCAR, and ECMWF had become convinced of the value of reanalysis for atmospheric research purposes and for the improvement of their own NWP skill. The adoption of reanalysis as a methodology enabled a huge number of new capabilities for NWP. It ensured the continuing integrity of the observational data archive—whose longevity was in grave danger due to dwindling curatorial resources—and led the publication of hundreds of scholarly papers that pushed the boundaries of our understanding of atmospheric dynamics and physics. All this can be ascribed to the wisdom of the NSF in seeing the value of such an enterprise.

Dynamical Seasonal Climate Prediction

As early as the late 1970s, I began to suspect that, despite the chaotic nature of day-to-day atmospheric fluctuations we normally ascribe to "weather," the Earth's climate might be predictable beyond the so-called deterministic limit due to slowly varying conditions and processes in the climate system. We began to explore this possibility with admittedly crude global atmospheric models at that time, and found encouragement in the relationship between variations in the sea surface temperature and the large-scale atmospheric circulation. Later, we found that there was a potential predictive relationship between land surface variations and the atmospheric circulation and precipitation on seasonal time scales. A firm scientific basis was needed for exploring and quantifying this seasonal predictability. Hand-in-hand with that academic question, there was a real nuts-and-bolts issue of how to exploit that predictability. At that time, seasonal predictions were made entirely on an empirical basis. We approached NSF, along with NOAA and NASA, to consider establishing a national research center to explore, understand, and quantify seasonal and longer time scale predictability and to develop the capability for regular dynamical seasonal prediction. Again, NSF took the lead, and the other agencies responded with enthusiastic support. After a rigorous peer-review process, the project was launched. The result of that and subsequent proposals was the establishment of the Center for Ocean-Land-Atmosphere Studies, which has thrived for over 20 years.

Ensuing research revealed great disparity among the results obtained by the various modeling groups that explored the question of seasonal prediction. As before, NSF recognized the importance of a national program to critically and quantitatively compare the various models and supported us in establishing the Dynamical Seasonal Prediction program. A sister program called PROVOST (PRediction Of climate Variations On Seasonal and inter-annual Timescales) was established in Europe. More recently, after the World Climate Research Program announced a new strategy for Coordinated Observation and Prediction of the Earth System (COPES), the Atmospheric Sciences Division of NSF has taken the leadership role in the United States to engage the scientific community and other federal agencies in serious discussions about this emerging strategy.

support to the Interagency Community Coordinated Modeling Center, where users can request specific model runs and visualize the results.

With the enormous successes have come significant challenges. More and more, numerical weather forecast models are assimilating increasing amounts of satellite data, with assimilation of radar data promising to improve short-term convective storm forecasting in the near future. Furthermore, use of numerous runs of the same model with slightly different initial conditions ("ensembles") or combining ensembles of runs with different models ("superensembles") are not only improving forecasts but also allowing an evaluation of model uncertainty and development of probabilistic predictions.

The climate challenge is even more significant, because simulating future climates needs to involve the interaction of the oceans, the Earth system, and the cryosphere, as well as the atmosphere and the Sun. Policymakers are demanding such runs be urgently produced at higher resolution than presently feasible (NRC, 1998a, 2001b), with effective horizontal grid spacing of 5 km or less. As in the case of weather forecasting, the use of ensembles and superensembles is essential to estimate uncertainty, and to generate probabilistic seasonal predictions and climate change predictions. Meeting the demand for improved climate modeling capability will likely require substantial increases in computational resources.

For several decades, the continuing rapid development of computer capability has enabled the ATM community to more or less meet its computing needs, but this seems no longer to be the case. Advancements in simulation, data assimilation, and prediction capabilities have in recent years begun to place serious demands on existing computational resources—demands that, increasingly, are not being matched by new investments. As pointed out in several reports (NRC, 2001b, 2004), this problem is growing, and is in fact more acute within the United States than in many other countries. For example, Japan, Germany, and the EU continue to make new investments in computing resources for the Earth sciences that match advancements in scientific capabilities.

Those considering the future needs of the community call for computers so large and with such significant cooling and power requirements that the decades-old solution of housing the NCAR supercomputer in the NCAR Mesa Laboratory is not going to be feasible in 10 years (Kellie, 2004). NSF, realizing the approaching challenge for the geosciences and other areas of science that rely on supercomputing, has issued a Request for Proposal (RFP) to support the development of a petascale computing environment (NSF, 2005b). UCAR coordinated a response (Ad Hoc Committee and Technical Working Group for a Petascale Collaboratory for the Geosciences, 2005). Also, while one can significantly increase computing power by combining multiple processors, the "latency" or time lag in communicating among the processors is a significant barrier to increasing computational speed (NRC, 2001b). In the meantime, through a contract with IBM, NCAR has worked hard to improve its computer infrastructure to keep up with the demand, and climate scientists in both NCAR and the university community have gained access to other computing facilities, including the Japanese Earth Simulator and the DOE Leadership-Class Computing Facility (LCF) at the Oak Ridge National Laboratory.

Computational scientists have joined in the development of community models, such as the Weather Research and Forecasting Model and the Community Climate System Model. As the scientific boundaries between atmospheric and geophysical and astrophysical sciences become more blurry, there will be benefits from pooling computing and intellectual resources. NRC (2001b) calls for national coordination in the form of "common modeling infrastructure" to facilitate model improvements and data formats to streamline the research process. NSF (2006) calls for a GEO computer and NRC (2005a) calls for multi-agency investment and minimizing barriers to international collaboration.

An additional challenge is storage and analysis of the enormous amount of data produced by the runs. Unless adequate storage is provided, in conjunction with computational capability to allow detailed analysis of the data, much of the research value is unrealized. NSF is to be commended for putting out an RFP for analysis of the products of recent climate assessment runs.

Looking to the future, it can be anticipated that the gap between research needs and available computational resources is going to become even wider unless action is taken to enhance such resources very substantially. NRC (2001b) makes the case that about 100–1000 times more computational capacity is required to meet existing needs than was available at the time. NRC (2004) emphasizes that a sustained program of support and development is needed to meet future needs. This issue is obviously larger than what can be addressed by ATM alone, and is arguably larger than what can be addressed by NSF alone. The committee concurs with the position taken by NRC (2004) that the government agencies that are the major users of supercomputing must take joint responsibility for the strength and continued evolution of supercomputing infrastructure in the United States, and that adequate and sustained funding must be allocated in the national budget.

It has long been recognized that strong computing facilities are of primary importance for advancing the atmospheric sciences. That need remains today and may be greater. How best to direct future investments in computing resources for the atmospheric sciences is a complicated issue that requires more careful study than possible in this report. Nonetheless, the committee is convinced that good science and important social impact would be enabled by better, faster models, which require more and more powerful computers. Supporting ever-larger and more capable computing infrastructure should be a high priority, but must be balanced by the other needs of the community, so as not to jeopardize maintaining observational facilities, and, especially, continued support in basic research. Meeting this demand will not likely be possible with the approaches used today and may require new organizational mechanisms, sources of funding, and partnering with other agencies, the private sector, or other nations.

4

Modes of Support and Key Activities

In this chapter, each of the major modes of support employed by the National Science Foundation (NSF) that now contribute specifically to the atmospheric sciences—that is, grants to individual and multiple Principal Investigators (PIs), small centers, large national centers, cooperative agreements to support facilities at universities and other locations, NSF-wide initiatives, interagency programs, and field programs—is described and their strengths and limitations are evaluated.

GRANTS

The Division of Atmospheric Sciences (ATM) supports academic atmospheric research principally through the proposal and peer review process for individual or multiple investigator grants. Among other activities, these grants support a large academic community of atmospheric scientists who pursue research that is essentially curiosity-driven. This basic research is fundamental to moving the field forward. Table 4-1 shows proposal statistics for ATM as compared to the Geosciences Directorate (GEO) as a whole and to the NSF averages. The bulk of the approximately 300 NSF-funded ATM grants each year are to individual PIs (in many cases with co-investigators), mostly at universities. The number of grants awarded each year has increased slowly over the past two decades (Figure 4-1), but there has been little trend over this time period in the success rate for grant proposals, which has fluctuated between approximately 40 and 50 percent for the division, despite increases in the number of proposals received (Figure 4-2). Most grants are for a three-year period.

TABLE 4-1 ATM Research Proposal Statistics for FY 2003

	ATM	GEO	NSF
Submitted proposals	~800	~4,000	~40,000
Competitive awards	~300	~1,500	~11,000
Average annual award (in 1996 dollars)	$127,000 ($108,300)	$147,000 ($125,350)	$136,000 ($116,000)
Average duration	3 years	3 years	3 years

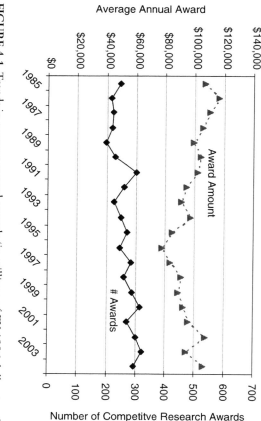

FIGURE 4-1 Trends in average annual awards (in millions of FY 1996 dollars) and number of grants awarded by ATM since 1985.

The average annual amount of ATM awards to PIs is about $127,000 per year, although actual support to an individual PI may be less if the grant is awarded to multiple investigators or more if allocations of computing or observing facilities are included in the award. For university faculty members, this amount normally includes up to two months of summer salary; support for graduate students, undergraduate students, or both; miscellaneous expenses such as travel, computing, and page charges; and institutionally determined fringe benefits and indirect costs. Over the past 10 years, 570 graduate students, on average, have been supported by ATM research grants each year, which is over half of the graduate students enrolled in atmospheric science departments (Jarvis Moyers, personal communication; NSF, 2006). The funding is committed for the duration of the grant, contingent on adequate progress being demonstrated though annual reports. Funding of investigators in nonacademic institutions proceeds similarly.

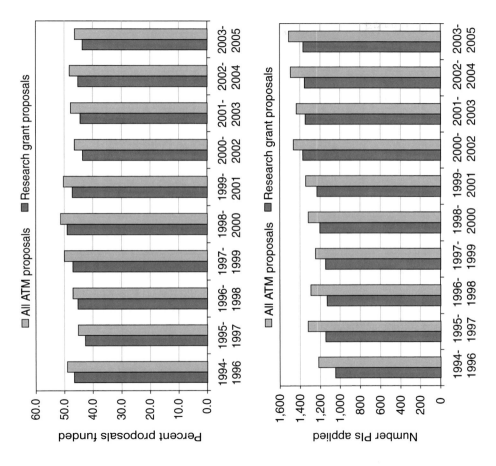

FIGURE 4-2 Top: Percent of proposals funded. Bottom: Number of proposals received.

Most grants are unsolicited; scientists with an idea for a research project send in a proposal which is then judged on the basis of scientific excellence and potential broader impacts, such as educational and other societal benefits. A small number of grants of limited scale and duration are awarded as part of the Small Grants for Exploratory Research (SGER) program, which is intended to promote investigation of more radical ideas. NSF and ATM also solicit proposals that address priority research areas or other specific objectives (e.g., Box 4-1). Often, these directed research programs respond to needs identified by the community, thereby alleviating the

BOX 4-1
Focused Programs That Are Community-Driven

Ongoing Programs with an Annual Competition for Funding:

Coupling, Energetics, and Dynamics of Atmospheric Regions (CEDAR) is a broad-based upper-atmospheric research program with the goal of understanding the behavior of atmospheric regions from the middle atmosphere upward through the thermosphere and ionosphere into the exosphere in terms of coupling, energetics, chemistry, and dynamics on regional and global scales.

The Geospace Environment Modeling (GEM) program supports basic research into the dynamical and structural properties of the magnetosphere. One of the objectives is the construction of a global geospace general circulation model with predictive capability.

Solar and Heliospheric Interaction (SHINE) research focuses on the connections between eruptive events and magnetic phenomena on the Sun and the corresponding solar wind structures in the inner heliosphere. The goal of SHINE research is to enhance both our physical understanding and predictive capabilities for solar-driven geoeffective events.

Earth System History (ESH) is a cross-divisional research program, which is managed by ATM's Paleoclimate Program Director. The program seeks to provide better understanding of Earth's paleoenvironmental system and its evolution over geologic time by (a) documenting the past temporal and spatial variability of the Earth system, (b) assessing the rates of change associated with this variability, and (c) determining the sensitivity of the Earth system to variations in climate-forcing factors. Note: Fiscal Year 2007 solicitation has been postponed while NSF reevaluates this program.

The Geoscience Education program aims at initiating or encouraging innovative geoscience education activities. It specifically seeks projects that are informed by results of current education-related research or that conduct educational research with a geoscience education venue.

The Opportunities for Enhancing Diversity in the Geosciences program supports activities that will increase the number of members of underrepresented groups that (a) are involved in formal precollege geoscience education programs; (b) pursue bachelor, master, and doctoral degrees in the geosciences; (c) enter geoscience careers; and (d) participate in informal geoscience education programs.

Recent Solicitations for Proposals on Targeted Topics:

The Pilot Climate Process and Modeling Teams (CPT) program was co-sponsored by NOAA and NSF. The goal was to further the development of global coupled climate models by enhancing collaborations between theoreticians, field observationalists, process modelers, and the large modeling centers.

The Water Cycle Research initiative was intended to enhance innovative basic research contributing to the understanding of the water cycle and its function as a transport agent for energy and mass (water and biologically/geochemically reactive substances).

concern that investigators must shoehorn their proposals to meet research priorities that do not necessarily reflect community goals. This mechanism is used more prominently by the upper atmospheric section.

There are several grant programs directed at young faculty and underrepresented groups. For example, the NSF-wide Faculty Early Career Development (CAREER) and the Presidential Early Career Awards for Scientists and Engineers grants target young, tenure-track faculty investigators who have not yet been awarded tenure. The number of these early career grant proposals is relatively small in ATM because of the relatively small number of tenure-track faculty in the field. GEO has grant programs that seek to enhance demographic diversity, including targeted programs for historically black colleges and universities, for tribal colleges and universities, and for improving female and minority representation.

While NSF grants from ATM are important for private-sector research companies, they are crucial to the career of university faculty members. The more mission-oriented agencies (e.g., National Aeronautics and Space Administration [NASA], National Oceanic and Atmospheric Administration [NOAA], Department of Energy [DOE], Environmental Protection Agency, Department of Defense, and the Federal Aviation Administration) support extramural research, but these funds are granted on the basis of mission relevance and scientific merit. Because NSF funding decisions are made primarily on the grounds of scientific excellence, there is a perception in some academic programs that success in obtaining NSF grants is considered more important to academic advancement.

Small science and technology oriented businesses can also apply for Small Business Innovation Research (SBIR) and Small Business Technology Transfer Research (STTR) grants through an NSF-wide solicitation each year (NRC, 2004). STTR projects must involve at least one small business and one not-for-profit research group, usually from an academic institution. SBIR and STTR grants, which receive about 2.7 percent of the NSF's extramural research budget, have funded the development and demonstration of a number of innovative instruments currently used in atmospheric research.

An increasing fraction of NSF grants are for multiple PIs collaborating on a larger-scale project (see Figure 4-3). In particular, multi-PI grants support modeling and measurement efforts. Atmospheric scientists have long recognized the value of collaboration (NAS/NRC, 1958) and are increasingly seeing the need to form teams that can access the multiple skills, tools, and facilities that are frequently required to plow new scientific ground. The demand on ATM for multi-investigator project funding is likely to continue to grow. An issue that arises as the scale grows is the ability for agencies to work together, and for agencies to coordinate with international partners, in the fostering and support of such programs.

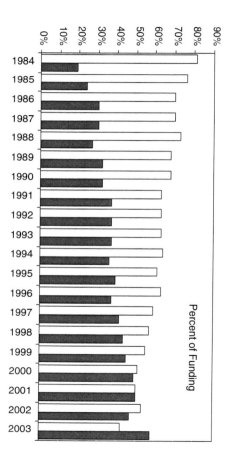

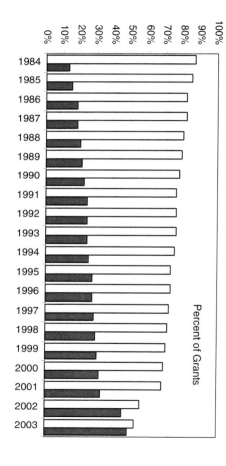

FIGURE 4-3 Percent of grants (top) and funding (bottom) awarded to single PIs (white) and multiple PIs (grey).

Increasingly, advances in modeling capabilities rest on critical collaborations and shared infrastructure. Likewise, the increasing complexity and frequent multidisciplinary nature of atmospheric science measurements—including laboratory experiments, ground-based and airborne field measurements, and advanced research instrument development and testing—often require collaboration of two or more research groups to be addressed effectively. Atmospheric field measurements often need to be performed at one or more remote sites, may require complex logistics involving site access or mobile measurement platforms, usually require the simultaneous measurement of multiple physical and/or chemical parameters, and normally require significant modeling capabilities for proper analysis. Another example where multi-PI grants have been effective is in support of global climate change at universities. An umbrella grant to Princeton University awarded during the 1980s–1990s was incredibly fruitful in fostering scientific advances in this complex, multidisciplinary area of research and in training the next generation of scientists. All of these factors push the requirement for multiple-PI projects.

There are often synergies between ATM PI grants and National Center for Atmospheric Research (NCAR) programs for both individual and multiple PIs. Many NSF grantees use research tools developed and maintained at NCAR. These include numerical models, equipment, and computing. Also, there is a great deal of science collaboration between NCAR scientists, who are frequently unfunded co-PIs on grants, and PIs from universities or the private sector in the conduct of their research, including field programs.

This mode of core grant support has benefited the atmospheric sciences in several ways. First, it has enabled a substantial volume of high-caliber scientific research. For example, grants to individual and multiple PIs have enabled the development of theory, analysis of observation and model results, process studies, provision of data to a broad suite of users, and development and acquisition of instruments by universities. Second, it has provided multiple options and flexibility in the ways ATM supports PIs, including unsolicited proposals, solicitation for new money that came in via various NSF-wide initiatives, ATM-initiated solicitations, and solicitations for field programs. This flexibility allows ATM to both encourage submission of proposals addressing focused topics or themes and to continually encourage PI-initiated proposals.

HIGH-RISK, POTENTIALLY TRANSFORMATIVE RESEARCH

Whereas other federal agencies, such as NASA, NOAA, and DOE, fund research directly related to their mission, NSF is unique in its flexibility to fund research without immediate mission application or need to have guarantee of success. Such curiosity-driven research is an important component of a thriving scientific field. An important type of this basic research is that which is particularly innovative or potentially high-risk. NSF is the primary place where scientists turn for support of this sort of research; other more mission-oriented agencies typically do not support it. Such research is instrumental in making major advances in the field and has the potential to be ground-breaking and lead to high payoff.

The NSF approach to reviewing and selecting research activities to support generally ensures that good science is funded and poor or mediocre science is not. In this process, the track record of the proposing investigator and likelihood of success of the proposed research are among the metrics used for evaluating proposals. These metrics may lead the PI community to put forward conservative proposals that produce incremental advances in atmospheric science. NSF program managers face a continual challenge to ensure that highly innovative, high-risk ideas are funded, since such proposals may have large potential payoffs. Such research efforts are more likely to fail, but may also lead to transformative discoveries.

Encouraging and identifying proposals that fall into this category and ensuring adequate support for them has presented challenges for NSF as a whole, despite the desire from NSF leadership to pursue innovation and risk taking (NAPA, 2004). As larger modes of support have expanded (e.g., small centers), the opportunities for such blue-sky funding is believed to have declined. Aside from those grants awarded through the SGER program, most proposals that might be considered high risk undergo the regular merit review process; thus it is unknown how much research of this sort is supported. Because peer reviewers tend to be risk averse, particularly innovative proposals may not fare well when competing against regular proposals. NAPA (2004) found that NSF's support for high-risk research could be enhanced by better communicating opportunities for such support to the scientific community, perhaps through specialized calls for proposals; by modifying the review criteria used to evaluate proposals to place more weight on innovation; or by subjecting high-risk proposals to a specially designated review process.

Currently, ATM does not set aside any funds specifically for high-risk research, but program officers are encouraged to be receptive to such proposals that come in through the regular grant process. In some cases, awards are made despite the lack of reviewer endorsement, shorter-duration proof-of-concept awards are made, or ATM or GEO reserves are used to fund such activities. One example of such an action by an ATM program director took place in the early 1980s when Dr. Ronald Taylor put funding into the newly emerging area of the MST (Mesosphere-Stratosphere-Troposphere) radar (see Case Study 5 in Chapter 2 of this report). This action accelerated progress in this field and now many such radars around the world are collecting valuable data. ATM does not track how many grants are awarded for high-risk proposals, either through the regular grant process or through the discretion of the program directors, or the outcomes of the high-risk research that is funded. Some high-risk projects that are of limited duration and of modest cost are supported through the SGER program. No more than 5 percent of any NSF program can be used for SGER awards; in ATM, typically 1 to 2 percent of each program's funds are applied to SGER. It is not entirely clear to investigators what funding mechanisms are available for support of high-risk projects that are larger in scope than that which an individual program director could fund. However, there is no analog to the SGER opportunity for potentially transformative research proposals whose cost exceeds the SGER threshold of $200,000.

SMALL CENTERS

Over the past two decades, NSF has begun to employ a small-center mode of funding. This mode was initiated by the Engineering Directorate, which introduced Engineering Research Centers (ERCs) in the early 1980s. Subsequently, the Office of Integrated Activities created Science and Technology Centers (STCs), which are designed to enable innovative research and education projects of national importance that require a center to achieve significant research, education, and knowledge-transfer goals shared by the partners. ERCs and STCs are funded for 10 years at the level of $2 million to $5 million per year. In addition, there are centers supported under the NSF-wide Information Technology Research (ITR) program and ATM supports some centers from core funds. Box 4-2 lists atmospheric science centers established over the past 15 years along with the science problems they are addressing. Because these centers are supported primarily by other parts of NSF, they provide an opportunity to expand the overall NSF level of support for atmospheric sciences.

The NSF Office of Integrative Activities currently supports 17 STCs. Two atmospheric-sciences-related STCs were awarded in the early years: the Center for the Analysis and Prediction of Storms (CAPS) housed at the University of Oklahoma and the Center for Clouds, Chemistry, and Climate (C4) at Scripps Institution of Oceanography. Although CAPS and C4 have been sunsetted as STCs, support for the research initiated at these centers has continued because of successful competition for ATM core funding. At present, ATM is represented by two STCs, the Center for Integrated Space-Weather Modeling (CISM) coordinated by Boston University and Center for Multi-Scale Modeling of Atmospheric Processes (CMMAP) based at Colorado State University.

The Division of Engineering Education and Centers currently supports 22 current ERCs. There have been a total of 46 centers since the program started in 1985, and the last competition for new centers was in 2006, with 5 funded. Currently, there is one ERC focused on atmospheric science research, the Center for Collaborative Adaptive Sensing of the Atmosphere (CASA).

The few atmospheric sciences STCs and ERCs—CAPS, C4, CISM, CASA, and CMMAP—have contributed or are currently contributing significantly in advancing innovation and research in the atmospheric sciences. The STC and ERC programs provide participating investigators with long-term, relatively stable funding of sufficient size to tackle large problems. They involve the creation of large, interdisciplinary research efforts with

BOX 4-2
Small Atmospheric Centers Supported by NSF

Center for Analysis and Prediction of Storms (CAPS) was an STC at the University of Oklahoma from 1989 to 2000, funded at a rate of $0.9 million to $1.5 million per year. The CAPS mission was the development of techniques for the computer-based prediction of high-impact local weather with operational Doppler radars serving as key data sources.

Center for Clouds, Chemistry, and Climate (C4) was an STC spearheaded by Scripps Institution of Oceanography from 1991 to 2001, funded at a rate of $1.5 million per year. The goal of C4 was to develop theoretical, observational, and modeling bases required to understand and predict Earth's changing climate as affected by clouds, radiation, and atmospheric chemistry and their interactions.

Center for Integrated Space-Weather Modeling (CISM) is an STC coordinated by Boston University, starting in 2002, funded at a rate of $4 million per year for up to 10 years. CISM consists of research groups at eight universities and several government and private nonprofit research organizations and commercial firms. The center's mandate is to construct a comprehensive physics-based numerical simulation model that describes the space environment from the Sun to the Earth, thus enabling reliable prediction of space weather events at least two days in advance.

Center for Multi-Scale Modeling of Atmospheric Processes (CMMAP) is an STC awarded in July 2006 to Colorado State University (CSU), funded at $19 million for the first five years. The primary objective of CMMAP will be to develop climate models with more accurate depictions of cloud processes, building on prototypes pioneered by researchers at CSU.

Center for Collaborative Adaptive Sensing of the Atmosphere (CASA) is an ERC led by the University of Massachusetts at Amherst, funded at a rate of $1.5 million to $2 million per year for up to 10 years. Established in late 2003, the center brings together a multidisciplinary group of engineers, computer scientists, meteorologists, sociologists, and government representatives to conduct fundamental

targeted goals. Such a goal-oriented research focus, with milestones and metrics, is a different environment than the work of the individual PI. Stable funding benefits graduate students and postdoctoral fellows, and allows researchers to focus on key science issues that extend beyond the regular grant cycle for single and multiple PIs. While the mandated 10-year lifetime of the centers may pose management challenges near the end of the award, it also forces the centers to maintain a relevant, cutting-edge research portfolio throughout their tenure. Indeed, each center in the atmospheric sciences that has already "graduated" has continued to operate in some form with research, develop enabling technology, and deploy prototype engineering systems based on a new paradigm: distributed collaborative adaptive sensing. These networks are deployed to overcome fundamental limitations of current tropospheric observational approaches by using large numbers of appropriately spaced sensors capable of high spatial and temporal resolution.

Linked Environments for Atmospheric Discovery (LEAD) is an ITR program led by the University of Oklahoma and established in 2003. It is funded at a rate of $11.25 million for five years. The transforming element of LEAD is dynamic workflow orchestration and data management, which will allow use of analysis tools, forecast models, and data repositories as dynamically adaptive, on-demand systems.

Global Multi-Scale Kinetic Simulations of the Earth's Magnetosphere Using Parallel Discrete Event Simulation is an ITR project at the Georgia Institute of Technology to develop scalable, parallel, numerical models for the simulation of space plasmas and the dynamics of the Earth's magnetosphere, based on Discrete Event Simulation (DES). The investigators will develop DES methods with situation-dependent physics, suitable for space physics problems, and then develop the algorithms required to execute these efficiently on massively parallel computer systems.

Environmental Molecular Sciences Institute (EMSI) at UC Irvine (AirUCI) is co-funded by the Chemistry Division and the Atmospheric Science Division, using the relatively new EMSI funding mode. AirUCI's research focuses on chemical reactions at air/condensed phase interfaces, an important emerging topic in atmospheric chemistry.

Tree-Ring Reconstruction of Asian Monsoon Climate Dynamics is a new five-year collaborative project at Columbia University. The project will use the science of dendro-chronology to examine the relationship between the Asian monsoon and the large-scale coupled processes that drive much of its variability.

additional funding coming from a variety of sources. In addition to their research objectives, STCs and ERCs have mandates to conduct education activities and to develop applications and knowledge transfer. The STCs and ERCs are required to spend approximately 20 percent of their resources on education and diversity programs, well beyond the requirements of other grants and agency requirements. Thus, the centers significantly broaden education resources. For example, CISM holds a two-week summer school that provides broad-based exposure to space weather in the entire Sun–Earth system, which has proved to be very successful (Simpson, 2004). ERCs are specifically mandated to include minority-serving institutions in the team. STCs and ERCs also have to devote considerable resources to knowledge transfer—making the products of the research useful to users in the real world. For ATM, this has meant moving atmospheric or space weather predictive capability from research into operations (NRC, 2000).

In addition to the well-established STC and ERC models, some NSF divisions are experimenting with other research center models. For instance, the Chemistry Division has established a series of Environmental Molecular Sciences Institutes (EMSIs). One of these, AirUCI centered at the University of California at Irvine (AirUCI, 2006) is co-funded by the Chemistry Division and the Atmospheric Science Division and focuses on chemical reactions at air/condensed phase interfaces, an important emerging topic in atmospheric chemistry. EMSIs are funded at about one-third the level of STCs or ERCs. AirUCI has six primary investigators from multiple departments at UC Irvine, as well as ten separately funded collaborators from three DOE National Laboratories, three senior collaborators from foreign institutions, and two community college faculty summer participants. One AirUCI investigator wrote the committee that the EMSI model is particularly effective in enabling a multidisciplinary research group to tackle significant problems in atmospheric science in a way individual investigators cannot, while maintaining a group structure that requires only modest overhead expenses. This relatively modest "small center" model may be very useful in dealing with other emerging atmospheric science topics, particularly multidisciplinary subjects that reach across NSF divisional structures.

In summary, these small centers have achieved their intended goals to: (1) support research and education of the highest quality; (2) exploit opportunities where the complexity of the research agenda requires the advantages of scope, scale, duration, equipment, and facilities, that a center can provide; and (3) support innovative frontier investigations at the interfaces of disciplines and fresh approaches within disciplines. Therefore, this research mode is effective in advancing the science and its transition to operation. All of these small atmospheric science centers have played pivotal roles in major scientific achievements in the field that led to direct societal benefits such as improved severe storm predictions or improved space-weather forecasting.

LARGE NATIONAL CENTER

One of the mechanisms used by NSF for support of research is a large national center. Typically designated as federally funded research and development centers (FFRDCs), they provide for a larger aggregation of research capability than that which could ordinarily be expected to occur at an individual university department. The largest NSF FFRDC is NCAR, located in Boulder, Colorado. The University Corporation for Atmospheric Research (UCAR), a nonprofit consortium of 70 North American universities with graduate programs in atmospheric sciences, has managed NCAR since its founding in 1960 through a cooperative agreement with ATM. This structure was designed to foster interactions and joint management between NCAR and the university community.

The specific objectives for NCAR were laid out in the 1959 "Blue Book" authored by the University Committee on Atmospheric Research ("UCAR"; see Box 4-3). The critical mass of resources that NCAR brings to bear on the atmospheric sciences includes computational resources, aircraft resources, observational capabilities, laboratories, and machine shops. An additional objective was to provide a personnel base that could support large-scale research, including interdisciplinary research. The center would have sufficient support personnel to enhance the research environment. The initial planning for NCAR called for half of the scientific staff to be from the atmospheric sciences with the remainder being from disciplines such as physics, mathematics, chemistry, and engineering. This disciplinary composition has evolved since 1959 as demanded by new research avenues in the atmospheric sciences.

Today, NCAR has about 220 scientists, 100 associate scientists, and 620 support personnel (which encompasses everything from software engineers to administrative assistants) who conduct research in the atmospheric and ocean sciences and in solar and space physics, and participate in a suite of activities that support the broad community. As shown in Box 4-4,

---

BOX 4-3

Four Compelling Reasons for Establishing a National Institute for Atmospheric Research identified in the "Blue Book" ("UCAR," 1959):

1. The need to mount an attack on the fundamental atmospheric problems on a scale commensurate with their global nature and importance.
2. The fact that the extent of such an attack requires facilities and technological assistance beyond those that can properly be made available at individual universities.
3. The fact that the difficulties of the problems are such that they require the best talents from various disciplines to be applied to them in a coordinated fashion, on a scale not feasible in a university department.
4. The fact that such an institute offers the possibility of preserving the natural alliance of research and education without unbalancing the university programs.

BOX 4-4
Overview of NCAR Organization, Activities, and Facilities

NCAR Organization:

Computational Information and Systems Laboratory, which houses the Institute for Mathematical Applications in the Geosciences and the Scientific Computing Division.

Earth Observing Laboratory (EOL), which includes the Atmospheric Technology Division (ATD) and the High-performance Instrumented Airborne Platform for Environmental Research (HIAPER); EOL maintains and deploys observational facilities for the lower-atmosphere research community.

Earth and Sun Systems Laboratory (ESSL), which houses much of NCAR's scientific research as well as its community models. ESSL includes

- Atmospheric Chemistry Division
- Climate and Global Dynamics Division
- High Altitude Observatory (HAO)
- Mesoscale and Microscale Meteorology Division
- The Institute for Integrative and Multidisciplinary Earth Studies (TIIMES)
- The NCAR library

Research Applications Laboratory (RAL), which includes the Research Applications Programs, is involved in a spectrum of activities relating to technology transfer and application of new knowledge to practical use.

Societal and Environmental Research and Education (SERE) Laboratory, including the Advanced Study Program, which offers postdoctoral positions that enable participants to explore the research areas of their choice, and the Institute for the Study of Society and Environment.

Strategic Initiatives are intended to bridge disciplines to advance Earth system science. Current initiatives include:

- Biogeosciences
- Community Spectro-Polarimetric Analysis Center
- Coronal Magnetic Fields
- Cyber infrastructure
- Data Assimilation
- Education and Outreach
- Measurement of Winds and Temperatures in the Upper Atmosphere
- Geographic Information Sciences
- Megacity Impacts on Regional and Global Environments: Mexico City Pollution Outflow Field Campaign (MIRAGE-MEX)
- Upper Troposphere-Lower Stratosphere

- Water Cycle Across Scales
- Weather and Climate Impact Assessment Science

NCAR Activities:

- Community model development, maintenance, support, analysis, and dissemination, e.g., Community Climate System Model (CCSM), Whole Atmosphere Chemistry Climate Model (WACCM), Weather Research and Forecast (WRF) Model
- Expensive large facility acquisition, maintenance, and support, e.g., aircraft, computers, Mauna Loa Solar Observatory (MLSO)
- Data storage and access
- Large field program logistical support in coordination with UCAR's Joint Office for Scientific Support (JOSS), part of which moved to NCAR in October 2005
- Long-term technology development, e.g., Cross Chain LORAN Atmospheric Sounding System, Global Positioning System (GPS), Lower Atmospheric Sounding System balloon soundings, Solo radar editing and analysis software, flux towers, eye-safe lidars, instruments to observe the Sun, and community instruments
- Virtual small centers to address larger interdisciplinary research questions, i.e., Strategic Initiatives listed above
- Major partner in support of small centers housed at universities, e.g., C4, CISM
- Intergovernmental Panel on Climate Change model runs

Lower-Atmosphere Facilities (EOL):

Aircraft

- HIAPER, high-altitude, long-range, high-performance Gulfstream V aircraft
- C-130, long-range, tropospheric, heavy-payload aircraft

Aircraft remote-sensing instrumentation

- ELDORA (ELectra DOppler RAdar), 3-cm high-resolution airborne Doppler radar, flown on a Naval Research Laboratory P-3 aircraft
- Airborne imaging microwave radiometer
- Multichannel cloud radiometer
- Scanning aerosol backscatter lidar

Ground-based remote sensing

- Raman-shifted eye-safe aerosol lidar
- S-Pol, S-Band Dual Polarization Doppler Radar

Surface and sounding systems

- Global Atmospheric Observing System (GAOS): Rawindsonde, housed in small trailer; employs GPS or LORAN-C navigation for winds
- Tethered Atmospheric Observing System: measurements on balloon tether
- Integrated Sounding System: GAOS, surface station, 915-MHz radar wind profiler, Radio Acoustic Sounding System (RASS) virtual-temperature profiler
- Multiple Antenna Profiler: enhanced 915-MHz radar wind profiler
- Integrated Surface Flux Facility: flux of sensible and latent heat, trace gases, and radiation; standard atmospheric and surface variables

continued

BOX 4-4 Continued

Solar Facilities (HAO):

The Mauna Loa Solar Observatory (MLSO) takes long-term synoptic observations of the Sun and makes the data available to a worldwide community. The instruments at MLSO include:

- Advanced Coronal Observing System, which consists of three instruments that monitor the flow of plasma and energy from the Sun's chromosphere through its corona and into interplanetary space

- Precision Solar Photometric Telescope, which measures brightness on the solar disc

- Experiment for Coordinated Helioseismic Observations (ECHO), in coordination with a second telescope operated by the Astronomical Institute of the Canaries at Tenerife, observes pulsations in the photosphere and low chromosphere, to monitor the Sun's energy budget in several important wavelength ranges

Advanced Stokes Polarimeter at the Dunn Solar Telescope at National Solar Observatory's Sacramento Peak site collects precise polarization measurements to infer the three-dimensional magnetic field and thermodynamic structure of the solar photosphere.

Fabry-Perot Interferometer at the Early Polar Cap Observatory at Resolute Bay measures wind speeds in the mesosphere, and will be used to support the Advanced Modular Incoherent Scatter Radar (AMISR), which will be deployed at Resolute in 2006.

NCAR and its scientists support the broad community in many ways, ranging from model development to maintenance of observing facilities and data archives. These scientists collaborate in large research programs involving many institutions as well as with scientists who visit NCAR through various fellowship programs. In the initial conception, NCAR was to be involved in only basic research in "recognition that there is need in atmospheric research for work to progress on a broader basis than that which is possible under the constraints imposed on applied research and development responsive to operational requirements" ("UCAR," 1959, p. 21). The programs at NCAR now include more applied research and transfer of the information, expertise, and technology developed to the public and private sectors; these efforts are often supported by other federal agencies. Indeed, about a third of NCAR funding comes from sources other than NSF (Figure 4-4). Furthermore, the recent undertaking of Strategic Initiatives, listed in Box 4-4, has

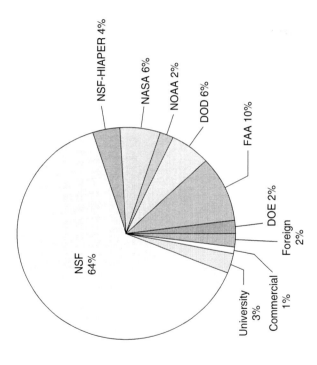

FIGURE 4-4 Sources of NCAR FY 2004 funding. Total funding was $138 million. Data provided by UCAR (Richard Anthes, UCAR, personal communication, July 22, 2005).

aimed to enhance interdisciplinary approaches to major research questions in the Earth system sciences.

UCAR is a not-for-profit consortium of 70 universities that grant doctorates in fields related to atmospheric science. At its inception, UCAR consisted of a president who oversaw NCAR with the help of a small staff and the advice of a Board of Trustees, who were elected from among the two member representatives from each of the UCAR universities. UCAR's primary activity is managing NCAR, but in the past few decades, UCAR has grown considerably, providing its own "national center" services, often in coordination with NCAR (see Box 4-5). In particular, UCAR supports its university members through the UCAR Office of Programs (UOP), which provides real-time weather data, digital library services, training to forecasters, field research support, and other activities. The yearly budget (December 31, 2004 figures) was $210 million for UCAR; 27 percent of which went to UOP and 71 percent to NCAR. UCAR currently employs 1,472 staff, of which 36 percent work directly for UCAR and 64 percent are at NCAR. NSF provides 65 percent of the funding for NCAR/UCAR.

BOX 4-5
UCAR-Activities Besides NCAR

The **UCAR Office of Programs**, whose portfolio includes

- Unidata, whose function is "providing data, tools, and community leadership for enhanced earth-system education and research"
- The Joint Office for Science Support (JOSS), which arranges logistics for international conferences and complex field programs, helps conduct the field program, and archives the field catalog and data (note that on October 1, 2005, part of JOSS will move from UCAR to NCAR)
- The COoperative Meteorological Education and Training (COMET) program, which trains forecasters from the National Weather Service, the military, and foreign weather services in the application of new research results and technology through face-to-face and distance-learning classes, in collaboration with University faculty and NCAR staff
- The Digital Laboratory for Earth Systems Education, which supplies datasets, imagery, and other educational resources to K-16 educators
- The National Science Digital Library, which is NSF's digital library for science, technology, engineering, and mathematics education
- The Constellation Observing System for Meteorology, Ionosphere, and Climate (COSMIC) program supports collection of meteorological data using the Global Position System (GPS) network of satellites
- The Global Learning Through Observations for the Benefit of the Environment program, an international inquiry-based education and science program to provide K-12 students with authentic science experiences through taking and analyzing environmental science measurements
- Visiting Scientist Program, which provides opportunities for scientists to visit other institutions

Education and Outreach, which supports

- Windows to the Universe Web site (UCAR, 2006b), which includes development of K-12 educational materials and professional development of K-12 educators
- SOARS, a multiyear program to entrain promising minority students into the atmospheric sciences, using mostly NCAR scientists as mentors

To some extent, it is difficult to differentiate between the roles of NCAR and UCAR. The two organizations work together to provide a range of activities, with staff and resources shared between them. For example, Unidata and JOSS provide significant university support, sometimes with the help of NCAR and university scientists, while the COoperative Meteorological Education and Training (COMET) program provides a venue for the NCAR and university communities to transfer new technology to operational forecasters. About 50 NCAR staff participate in UCAR's Scientific Opportunities in Atmospheric and Related Sciences (SOARS®) program each year, acting as scientific, writing, or community mentors.

In the opinion of the committee, NCAR/UCAR has been a highly successful center in terms of advancing knowledge of atmospheric science and providing community-based resources. NCAR/UCAR has met many of the objectives laid out in the Blue Book. Since the late 1950s, the atmospheric research enterprise has greatly expanded to its present state where impressive research capabilities exist in the universities, the private sector, and in federal laboratories. Even so, the fundamental rationale for a large national atmospheric sciences center outlined in the Blue Book still remains valid. The national center continues to serve important objectives of the atmospheric sciences community, as articulated in its stated vision:

"It is NCAR's mission to plan, organize, and conduct atmospheric and related research programs in collaboration with the universities and other institutions, to provide state-of-the-art research tools and facilities to the atmospheric sciences community, to support and enhance university atmospheric science education, and to facilitate the transfer of technology to both the public and private sectors."

The capabilities of each sector have increased tremendously since that time as have the myriad challenges and opportunities in the atmospheric sciences and allied fields. Thus, the challenge will be to prioritize and direct the activities of the large center so that it, together with the other research sectors in the atmospheric sciences, can best advance the field to the benefit of society. In making choices for allocating their resources, the large national center should continue to be guided by the following mandates in consultation with representative input from the broad U.S. atmospheric research community. It should

1. tackle large, complex research problems, in coordination with the universities, other federal agencies, and the private sector;

2. maintain standards of scientific excellence and openness that are commensurate with its university-based mission;

3. assume a share of the leadership in the atmospheric sciences community;

4. provide leadership in supercomputing in support of the atmospheric sciences and the modeling of the Earth system;

5. develop community models for climate, weather, space weather, and atmospheric chemistry;

6. develop advanced computational and numerical techniques and tools for use in atmospheric science;

7. enable field campaigns by coordinating their planning, managing, and logistics;

8. provide state-of-the-art archiving, access, analysis, and visualization tools for community datasets and data from NSF-sponsored field programs;

9. design, develop, and maintain state-of-the-art atmospheric instrumentation and observing platforms;

10. support education and diversity in the atmospheric sciences research community;

11. maintain vibrant postdoctoral and scientific exchange programs; and

12. foster opportunities to transfer knowledge and technology to public- and private-sector users.

These mandates are broadly in agreement with NCAR's existing mission. Over the past decades the atmospheric sciences community has thought a lot about the role of the large national center and these mandates attempt to encapsulate the collective view about what its goals should be.

Nevertheless, it is prudent to mention several potential challenges to such institutions. A key challenge is the establishment and communication of clear mechanisms for setting priorities in new directions as the center evolves to meet new research needs while continuing to ensure that it is meeting the needs of the nation in the purpose for which it was intended. A related challenge is the tendency for institutes such as NCAR to grow over time. Some have questioned whether NCAR/UCAR has become too large, perhaps at the expense of NSF-supported university research. The committee notes that NSF's level of support for NCAR over the past 30 years closely tracks that for PIs (see Figure 3-1). In fact, the percentage of the ATM budget spent on both NCAR and PIs has decreased as more resources have been devoted to observing and other facilities.

Maintaining an effective and balanced relationship with the university community may be the most significant challenge for NCAR. The center has a long track record of successful collaboration with university scientists to make progress on large scientific problems that are beyond the reach of a single university department or private-sector laboratory. This is consistent with the vision expressed in the Blue Book ("UCAR," 1959). These collaborations have originated in several ways, including through (a) scientist-to-scientist interactions, (b) large national or international programs (e.g., Global Atmospheric Research Program [GARP] Atlantic Tropical Experiment [GATE]), (c) NCAR initiatives (e.g., International H$_2$O Project [IHOP_2002]), (d) STC proposals, and (e) the development of large numerical models (e.g., Community Climate System Model). Through these collaborations, U.S. atmospheric research and operations have benefited greatly from the existence and productivity of NCAR.

Yet, scientific collaborations among widely dispersed investigators with different sets of priorities at other organizations are difficult to implement.

It can be easier to assemble most of the experts that are needed into one organization and to include only a few investigators from other organizations when necessary for specific research projects. In the current NCAR, there are elements that are truly collaborative with the university community, but there are also elements that are competitive with the university community. NCAR, the university, and the private-sector research communities have become so large and complex that there are new challenges for the center in terms of maintaining a balance between inward- and outward-looking efforts and in engaging a larger, more fragmented university and private-sector research community. Effective research partnerships, in the end, require that participants see mutual benefit with the partnership. But, it is also important that individual NCAR scientists perceive that NCAR management puts high value on collaborations with researchers outside of the center. Such partnerships are an important way that NCAR provides resources to the larger atmospheric science community. New ways to stimulate NCAR partnerships with the university and private-sector research community may be necessary.

Whereas there are many opportunities for collaboration between NCAR and university or private-sector scientists, decisions about NCAR strategic initiatives (e.g., recent new efforts in biogeosciences and water) could benefit from broader community input. In particular, new interactions could be instrumental in developing an agenda for the center that meets the needs and interests of both the large, and highly competent, in-house scientific staff and the broader atmospheric research community. Collaborations between large national centers (both existing and emerging) and university or private-sector scientists could be enhanced by new mechanisms to stimulate joint research initiatives at a larger scale than existing ad hoc collaborations.

COOPERATIVE AGREEMENTS TO SUPPORT OBSERVING FACILITIES

In addition to the facilities at NCAR, ATM uses cooperative agreements to support several facilities, often operated by universities and used by the broader atmospheric sciences research community (Table 4-2). These facilities provide scientists with instrumentation necessary to conduct cutting-edge science, are frequently utilized in field programs, and serve to meet educational objectives. For example, the CHILL Radar, which is operated by teams from both the Departments of Atmospheric Sciences and Electrical Engineering at CSU and has brought radar technology to the forefront, is available to the broad community by both onsite and remote control. Furthermore, CHILL staff members conduct training programs for students. Facility funding is provided through cooperative agreements

TABLE 4-2 Facilities Supported by ATM and Operated by Universities or Other Entities

Operational	Under Development
Lower Atmospheric Facilities:	*Lower and Upper Atmospheric Facility:*

Lower Atmospheric Facilities:

The **CHILL Radar**, operated by Colorado State University, is a deployable dual Doppler radar. It provides remote-sensing data of the lower atmosphere in support of collaborative radar research with federal, state, and academic research entities, and the meteorological community.

The **King Air Aircraft**, operated by the University of Wyoming, has been highly modified to support atmospheric and remote sensing instrumentation and is used to obtain in situ and remotely sensed atmospheric measurements of the lower atmosphere.

Upper Atmospheric Facilities:

SuperDARN Radar Network, operated by Johns Hopkins University Applied Physics Laboratory and the University of Alaska, is located in Canada and Alaska, and is part of a larger international network of sites. Its observations contribute to the global specification of the ionospheric electric potential pattern.

Four large incoherent-scatter radar facilities located along a longitudinal chain from Greenland to Peru:

Sondrestrom Radar Facility operated by SRI International is the northernmost radar in the chain, located in Kangerlussuaq, Greenland. It is used to further the understanding of the high-latitude upper atmosphere and space environment, in particular by investigating polar magnetosphere-ionosphere coupling under varying solar forcing conditions.

Lower and Upper Atmospheric Facility:

The **Constellation Observing System for Meteorology, Ionosphere, and Climate (COSMIC)** is being built through a partnership between NSF, NASA, NOAA, the U.S. Air Force and U.S. Navy, and Taiwan. COSMIC will include a fleet of six low-Earth-orbiting satellites to measure the refraction, retardation, and bending by Earth's atmosphere of radio waves transmitted by the fleet of 28 DoD supported high-Earth-orbit GPS satellites. The refraction of the radio waves yields a measure of electron density in the ionosphere and density variations in the stratosphere and troposphere, which in turn yield vertical profiles of temperature, water vapor, and pressure. COSMIC was successfully launched April 15, 2006.

Upper Atmospheric Facility:

Advanced Modular Incoherent Scatter Radar (AMISR) is a modular, mobile radar facility for studying the upper atmosphere and observing space weather events. SRI International is leading the development and construction of AMISR along with several other partners.

continued

TABLE 4-2 Continued

Operational	Under Development
Millstone Hill Radar, operated by the Massachusetts Institute for Technology, is located outside of Boston, Massachusetts, and is used to investigate mid-latitude magnetosphere, magnetospheric-ionospheric coupling, and thermospheric-ionospheric processes.	
Arecibo Observatory, operated by Cornell University, is located in the Karst region of Puerto Rico and explores the mesosphere, thermosphere, and the F region energetics and dynamics, as well as ionospheric-thermospheric coupling. It is particularly well suited for studies of the topside ionosphere.	
Jicamarca Radio Observatory, operated by Cornell University, is located on the magnetic equator near Lima, Peru. This instrument examines topside light ion distribution, latitudinal variability, and storm time response, F region thermal balance, and E region composition profiles of the equatorial region.	

with NCAR and a number of universities, to acquire, maintain, and operate specific observational and cyber infrastructure facilities or services that support the research and educational activities of NSF-sponsored projects, scientists, and students.

New emerging modes are under consideration for supporting facilities. In FY 2002, ATM created a "midsize facility" account to enable construction of new infrastructure that did not meet the minimum cost consideration for the NSF-wide Major Research Equipment and Facilities Construction account line (about $75 million for GEO), but costs in excess of the resources of any individual ATM program or section. The first two projects to be supported by this account are the AMISR and the Constellation Observing System for Meteorology, Ionosphere, and Climate (COSMIC). AMISR has just been initiated; the grant for building it was awarded to SRI International in Menlo Park, California. COSMIC is being operated by UCAR. COSMIC is a large effort in which ATM is but one of the players.

A number of issues arise in making choices about which observing facilities to support and how to implement them. One must consider the balance in needs for observational platforms across the disciplines (i.e., climate, mesoscale convection, space weather, etc.) and in needs for different types of platforms (i.e., aircraft, radars, etc.). After their scientific utility is established, it is not clear which factors and weightings should determine the distribution of investments in observational platforms; obvious considerations include the number of researchers seeking access and the capital and maintenance costs of the facilities, but other priorities may also apply. Another dimension of balance to consider is the extent to which small or large centers, universities, or private-sector entities should support development and maintenance of observational platforms. Similarly, NSF must determine an appropriate balance for maintaining and keeping existing facilities up to date, retiring facilities as appropriate, and developing new facilities. Since some facilities are very expensive to operate and maintain, it is important that NSF frequently and carefully continue to determine which facilities are essential and which can be phased out with least scientific impact. How best to utilize partnerships of NSF with other agencies that support observational facilities is another area of consideration. NSF has collaborated with other agencies to develop observing facilities, as it is currently doing in the case of COSMIC, and to deploy observing facilities for large field programs, such as the INdian Ocean EXperiment (INDOEX) campaign. There may be further opportunities to build such collaborations.

NSF-WIDE INITIATIVES

ATM participates in a number of NSF-wide, interagency, and international programs, which in some cases require different approaches to providing support. The NSF-wide emphasis areas result from national initiatives spearheaded by Congress or the President, or else are activities such as the STCs that NSF leadership chooses as a priority. They can bring new funds into the Foundation, which are then distributed to relevant divisions. Since 2000, ATM has received additional funds toward five NSF priority areas, as well as from the STC and ERC programs described previously (Table 4-3). Typically, these funds are distributed as grants to individual areas, as well as from the STC and ERC programs described previously and multiple PIs who respond to specialized calls for proposals.

FIELD PROGRAMS

Organized field programs that provide atmospheric observations designed to study specific processes continue to be integral to atmospheric research. Major field programs supported by ATM during the past decade are described in Table 4-4. Field programs are supported through a combination of modes, usually including grants to individuals or groups, facilities from NCAR or universities, NCAR field support, and often involve

TABLE 4-3 Investments in ATM Research from NSF-wide Priority Areas (in millions of dollars for each fiscal year)

Priority Area	2000	2001	2002	2003	2004
Biocomplexity in the Environment: improve environmental forecasting capabilities; enhance decision-making tools; and integrate human, social, and ecological factors into investigations of the physical environment and environmental engineering.	0.00	7.50	7.40	7.40	12.00
Information Technology Research: deepen fundamental research on large-scale networks and create new integrative software and advanced architectures for high-end computing.	0.00	3.40	3.40	4.60	5.00
Nanoscale Science and Engineering: develop and strengthen promising fields (including nanobiotechnology, manufacturing at the nanoscale) and establish the science and engineering infrastructure and workforce needed to exploit new capabilities in systematic organization, manipulation, and control of matter at atomic, molecular, and supramolecular levels. NSF activities are part of the larger, cross-agency National Nanotechnology Initiative.	0.00	0.00	0.50	0.50	0.60
Mathematical Sciences: deepen support for fundamental research in the mathematical sciences and statistics and integrate mathematical and statistical research and education across the full range of science and engineering disciplines.	0.00	0.00	0.00	1.50	2.40
Human and Social Dynamics: draw on recent convergence of research in biology, engineering, information technology, and cognitive science to investigate the causes and ramifications of change and its complex consequences—cultural, economic, individual, political, and social.	0.00	0.00	0.00	0.00	0.50

TABLE 4-4 Recent Large ATM Field Projects (over $1 million in facility deployment costs)

Description of Field Program	Estimated Support from NSF Grants
The **first Aerosol Characterization Experiment** (ACE-1) in FY 1995 was the first of several experiments to characterize the chemical and physical processes controlling the evolution and properties of atmospheric aerosols and radiative climate forcing. NOAA and Australia also provided facilities.	$5.0 million

continued

TABLE 4-4 Continued

Description of Field Program	Estimated Support from NSF Grants
The **Surface Heat Budget of the Arctic Ocean (SHEBA)** in FY 1998 was a multi-agency program supported by NSF's Arctic System Science Program. Its goal was to acquire data on pack ice that covers the surface of the Arctic Ocean. The study involved many research facilities, including ones from DOE, the Office of Naval Research, and Japan.	$15.0 million
In FY 1999 the **Indian Ocean Experiment (INDOEX)** addressed natural and anthropogenic climate forcing by aerosols and feedbacks on regional and global climate. Participants contributed research facilities from U.S. agencies, Europe, India, and island countries in the Indian Ocean.	$5.0 million
The **Mesoscale Alpine Experiment (MAP)** was an FY 1999 coordinated international effort to explore the three-dimensional effects of complex topography. The goal was to combine advances in numerical modeling with those in remote observing technology. Researchers and facilities from 12 countries were active participants. NOAA and several countries also provided research facilities.	$7.5 million
Tropospheric Ozone Production About the Spring Equinox (TOPSE) was an FY 2000 study that investigated the chemical and dynamical evolution of tropospheric chemical composition over continental North America during the winter-to-spring transition. Ozone budget, distribution of radical species, sources and portioning of nitrogen compounds, and composition of volatile organic carbon species were determined. NASA, Canada, and numerous universities provided research facilities.	$2.8 million
Eastern Pacific Investigation of Climate (EPIC) was conducted in FY 2001 to address processes that determine the nature of deep convection in and near the East Pacific Intertropical Convergence Zone; the evolution of the vertical structure of the atmospheric boundary layer; and how sea-air coupling affects ocean mixed-layer dynamics and sea surface temperature in the East Pacific warm pool. NOAA and Mexico also provided research facilities.	$5.5 million
ACE-Asia, conducted in FY 2001, focused on climate forcing caused by aerosols over eastern Asia and developed a quantitative understanding of the gas/aerosol particle/cloud system. NASA, NOAA, DOE, the U.S. Navy, Australia, Japan, China, France, the United Kingdom, and Korea also provided research facilities.	$8.0 million
The **Maui Mesosphere and Lower Thermosphere (MALT)** campaign started in FY 2001 and continues today. It is using nested instrumentation with the 3.7-meter-diameter telescope at the Maui Space Surveillance Complex to study dynamical coupling between the mesosphere and the lower thermosphere. The Air Force Office of Scientific Research also supports this field campaign.	In FY 2005, 5 awards and 1 supplement totaling ~$1 million

continued

TABLE 4-4 Continued

Description of Field Program	Estimated Support from NSF Grants
The **International H₂O Project (IHOP_2002)** in FY 2002 examined the moisture tracks that fuel large convective storms in the Midwest, to better understand when and where these massive storms form and how intense they will be. NOAA, NASA, France, and Germany provided research facilities.	$6.4 million
Bow Echo and MCV Experiment (BAMEX) in FY 2003 studied the life cycles of mesoscale convective storm systems. The study combined two related programs to investigate bow echoes, especially those that produce damaging winds, and larger convective systems that produce long-lived mesoscale convective vortices. NOAA and Germany also contributed research facilities.	$3.6 million
The **North American Monsoon Experiment (NAME)**, an FY 2004 joint Climate Variability and Change (CLIVAR) and Global Energy and Water Cycle Experiment (GEWEX) project, was aimed at determining the sources and limits of predictability of warm-season precipitation over North America. The project focused on the key components of the North American monsoon system and its variability within the context of the evolving land surface-atmosphere-ocean annual cycle. NOAA and Mexico also contributed research facilities.	$3.6 million
The **Rain in Cumulus over the Oceans (RICO)** project was completed in January 2005. Its objective was to characterize and understand the properties of trade-wind cumulus clouds at all spatial scales, with special emphasis on determining the importance of precipitation. University of Wyoming provided research facilities.	$3.8 million
The **Terrain-induced Rotor Experiment (T-REX)** is the second phase of a coordinated effort to explore the structure and evolution of atmospheric rotors (intense low-level horizontal vorticies that form along an axis parallel to, and downstream of, a mountain ridge crest) as well as associated phenomena in complex terrain. The initial, exploratory, phase of this effort, the Sierra Rotors Project, took place in early spring 2004 in Owens Valley, California; T-REX was conducted in the same location in March and April 2006. The campaign utilized the HIAPER and King Air aircraft supported by NSF and the United Kingdom contributed the BAe146 aircraft.	$1.25 million
The **Megacities Impact on Regional And Global Environment—Mexico (MIRAGE-Mex)** field campaign took place in March 2006 and examined the chemical and physical transformations of gases and aerosols in the polluted outflow from Mexico City. The campaign brought together observations from ground stations, aircraft, and satellites. MIRAGE-Mex was organized by NCAR-ACD on behalf of the atmospheric sciences community and included support from NOAA, DOE, and Mexico.	$2.2 million

other agencies or countries. ATM supports smaller field programs through individual investigator grants and the facilities deployment pool. However, ATM supports large field programs in a variety of ways: as the lead agency (e.g., Bow Echo and MCV Experiment [BAMEX], IHOP_2002), as a major partner in an international effort (e.g., Tropical Ocean and Global Atmosphere Coupled Ocean Atmosphere Response Experiment [TOGA COARE]), as a supporting agency for field programs sponsored by other agencies (e.g., Boreal Ecosystem-Atmosphere Study [BOREAS], led by NASA), and, on occasion, supplying NSF facilities for which other agencies pay. NSF-funded PIs can also participate in field campaigns sponsored by other agencies through individual grants. ATM indirectly supports field programs by supporting investigators to develop research capabilities that are then employed in campaigns funded by other agencies. In the case of INDOEX, the C4 STC was instrumental in initiating and carrying out the field program. To facilitate the planning of field programs, ATM requires those interested in using facilities from the NSF deployment pool to submit requests as much as two years in advance. The procedures for reviewing field programs were updated in February 2005 (NSF, 2005a).

Field programs are expensive in terms of financial, facility, and personnel cost. In addition to immediate scientific results, they produce a wealth of data that can be mined repeatedly as new questions emerge as the field matures, because an area is revisited, or if there is a need for an expanded sample. In addition, they allow the exploration of phenomena in a wealth of climate regimes and geographical locations. As the atmospheric sciences have become more complex, conducting field programs has presented new challenges for ATM in determining how to support these efforts, including:

1. *Increased demand for facilities.* Particularly for the large and diverse lower-atmosphere community, there is significant demand for facilities that often leads to conflicts in scheduling. Carefully developed protocols for requesting facilities years in advance, negotiation with NSF program officers and facility providers, and input from the Observing Facility Advisory Panel have often, but not always, resolved conflicts. The problem is exacerbated by the fact that scheduling is often driven by probable weather and the scheduling of other facilities belonging to other agencies and countries (e.g., University-National Oceanographic Laboratory System) or the schedules of cooperating institutions.

2. *Need for a strategic approach to data archiving.* Access can be challenging for those desiring data from operational observational and monitoring networks (including surface, upper air, radar, and satellite), as well as from field-program data, historical data, and numerical model data. Currently, there are varied destinations for data archival, including

NCAR, Web sites set up by universities, and data archives established by other government agencies (e.g., National Climatic Data Center). For lower-atmospheric field campaigns back to the early 1990s, UCAR/JOSS has served as a center for data archiving for observational data and model simulations, or as a clearinghouse for PI-supported datasets archived elsewhere. In addition, the NCAR Research Aviation Facility has some aircraft data archived back to the 1980s, and GATE data are archived on the NCAR Mass Store, with hard copy and microfilm in the UCAR Archives. Likewise, HAO maintains data archives from its solar instruments. Furthermore, the Coupling of Energetics and Dynamics of Atmospheric Regions, a program addressing issues related to the upper atmosphere, has maintained an archive at NCAR for over a decade; and the solar physics community makes data on the Sun available on the Web through the interagency and international Virtual Solar Observatory.

Other government agencies, such as NASA, NOAA, and DOE, also have made efforts to establish data archives for data from field programs, satellite instruments, and monitoring networks. For example, NASA operates nine Distributed Active Archive Centers that provide storage and access to a wide range of environmental observational and model data (NRC, 1999c). For NSF-funded research, there is not always a clear responsibility for providing archived data to researchers from both large, multi-investigator field experiments and small field experiments, and the decisions made vary from experiment to experiment. Thus, data archival formats, quality control, and metadata are not necessarily standardized. The older datasets are in formats that are increasingly inaccessible, and some data from earlier but potentially significant field programs are not archived in a central place. Further, from time to time, retiring scientists are approaching JOSS or UCAR, wishing to find a permanent home for potentially valuable photographs, notes, or data. Finally, there are datasets residing in the NCAR archives that are valuable but difficult to use (old photographic plates of solar images from the HAO Climax Observatory, films from GATE and MONEX). At present, there is neither a formal procedure nor is funding set aside for dealing with these problems. However, JOSS and the NCAR Archives have worked with the researchers on an ad hoc basis to ensure that valuable resources and datasets are not lost, and to help access historic data.

3. *Maintaining access to data analysis tools.* It is becoming increasingly difficult to access older data from the standard observing network and from field programs because changing technology and analysis packages make these datasets more difficult to analyze. Even when the data are readily available, there are no standardized plotting/analysis software packages available. To be able to compare analyses from different cases, it is useful to be able to look at data plotted and analyzed with the same software package.

4. *Supporting data analysis.* Inadequate time and resources for analysis of data collected in the field has been a problem for decades. LeMone (1983) reported that it took six years to reach the peak in publications from GATE data. There was a time lag of five to six years between the Cooperative Convective Precipitation Experiment (1981) and the peak in resulting publications, and the peak in Florida Area Cumulus Experiment publications was four years after the experiment. Some scientists analyzing TOGA COARE data ran out of funding before they completed analysis and publication; some even ran out of funding before they obtained all their data.

A two- to ten-year post-analysis phase is recognized in the lifetime of a generic large NSF field program, discussed in the recently released document, "Field Program Support at UCAR" (UCAR, 2005). However, NSF's new procedures for reviewing field programs (NSF, 2005a) emphasize advance notice more than the post-field phase. Because analysis comes at the end of a field program and competes against the start of other new field programs, it is at times subject to reduction in support. Thus, support for field data archives, visualization tool development, and analysis is not commensurate with the investment in obtaining the measurements and the full benefit from the investment in a field program often is not realized. Providing adequate time for careful analysis and synthesis of field data, which today typically involves complementary numerical simulations, increases the probability of significant payoff. Grant durations longer than three years allow more time for data analysis.

5. *Spacing of field programs.* Increasing the time between field programs allows more time and money for data analysis, and could mitigate the increased demand for facilities. However, these factors have to be balanced against the benefits of more closely spaced field programs. The large infrastructure maintained to operate the facilities requires a certain level and frequency of use, not only to justify its existence, but to test instruments and maintain proficiency of the personnel, a requirement for airplane pilots. Furthermore, field programs are effective ways to inspire and recruit new students and to stimulate new questions.

6. *A need for longer-term sustained intensive measurements.* While ATM has a distinguished record in supporting long-term measurements of the upper atmosphere (Table 4-2) and the Sun (Box 4-4), current ATM policies and procedures for lower-atmosphere field programs are consistent with instrument deployments of the order of a few months. However, many problems related to weather and climate—for example, the interaction between the atmosphere and Earth's surface in the context of heat, moisture, or biogeochemical cycles—require sustained, specially designed, and focused measurements for a complete annual cycle or even several years. There are examples where ATM supported longer-term measurement goals

by supporting field programs on an episodic basis (e.g., First ISLSCP [International Satellite Land Surface Climatology Project] Field Experiment in the 1980s), but sustained measurements are often needed. There are also efforts within other divisions of NSF to develop capabilities for long-term observations over the ocean (e.g., Ocean Research Interactive Observatory Networks Ocean Observing Initiative [ORION OOI]) and the land surface (e.g., the proposed Consortium of Universities for the Advancement of Hydrologic Science, Inc. [CUAHSI] Hydrological Observatories, Long Term Ecological Research). Operational weather- and climate-monitoring networks provide observations over the longer term, but often not at the intensive level needed for process studies. ATM has not yet clearly articulated mechanisms for supporting field programs that require continuous, longer-term (i.e., up to multiyear) deployment and observations not available from operational monitoring networks. This type of observation protocol is generally ill suited to the existing funding opportunities, in part because they were prohibitively expensive until recently. Many instruments that would be used are now less expensive, making it reasonable to deploy them in the field for longer durations.

7. *Adapting to a changing international scene.* Historically, the United States usually has been the leader or at least a major partner in international field efforts. In the past few years, however, the major leadership in field programs has started to come from other nations. For example, the African Monsoon Multiscale Analysis field program is a large international field program supported by the European Union and led by France.

8. *Development of innovative observing techniques and methods.* For the U.S. atmospheric science community to remain at the cutting edge of field research, innovative techniques and methods need to be developed in order to obtain the observations needed to test hypotheses, better resolve the variability and structure of the atmosphere, and understand the coupling of the atmosphere to the land, cryosphere, ocean, and space. Once developed and proven these new methods need be transferred to facilities that can make them available to the broader community.

EDUCATIONAL ACTIVITIES

Each mode of support employed by ATM provides some resources for educational activities (see Table 4-5). Most of ATM's support of science education is accomplished through traditional research grants, which allow undergraduate and graduate students and postdoctoral scientists to participate in research efforts directly. At some universities, ATM has awarded block or umbrella grants, renewed every 3–5 years, that provided the advantage of a clear funding track for students throughout their Ph.D. tenure. NSF-wide and ATM-led initiatives also support a wide range of

TABLE 4-5 Examples of Educational Activities Conducted Using Each Mode of Support

Mode of Support	Educational Activities
Single and multiple PIs	• Undergraduate and graduate student research through research grants • Postdoctoral research through research grants • REUs as separate PI-funded activity
Small centers	• Undergraduate and graduate student research • Postdoctoral research • Community education resources (e.g., CISM summer school) • Graduate student communities and mentoring • K-12 science education • Informal science education • Undergraduate education and course development
Large center (NCAR/UCAR)	• Advanced Study Program for postdoctoral researchers • Young Faculty Forum • SOARS® • Resources for graduate students • Community-wide summer workshops • Meeting for heads and chairs of UCAR member departments • Visiting Scientist program • Sabbaticals from teaching • Cooperative Meteorological Education and Training (COMET) (NOAA, Navy, Airforce, Meteorological Service of Canada) • Numerous projects through UCAR Education and Outreach (funded by NASA) • Summer colloquium for graduate students and postdoctoral researchers
Cooperative agreements for university and other facilities	• Provide facility for graduate and undergraduate research • Provide venue for REU programs (MIT Haystack, Arecibo, CHILL Radar) • Make data available via the Web (e.g., radar data)
NSF-wide initiatives	• Provide resources for graduate research • Provide geoscience diversity initiative funded programs at a professional society (AMS) and a facility (Arecibo)
Interagency programs	• Provide resources for graduate research, postdoctoral fellowships, sabbatical and scholarly exchanges
International collaboration	• Provide resources for graduate research, postdoctoral fellowships, sabbatical and scholarly exchanges

other educational activities. At the NSF-wide level, the Research Experiences for Undergraduates (REU) program provides support for undergraduates in individual projects as well as special REU summer-site programs. NSF supports graduate students through the NSF Graduate Research Fellowship Program. ATM also provides scholarships through the American Meteorological Society and postdoctoral fellowships through UCAR. A number of educational efforts are organized through UCAR and NCAR. A prime example is the effort to bring underrepresented minorities into the atmospheric sciences through the Scientific Opportunities in Atmospheric and Related Sciences (SOARS®) program. SOARS® has been successful at increasing the participation of African American, American Indian, and Hispanic/Latino students enrolled in master's and doctoral degree programs in the atmospheric and related sciences (see Box 4-6). ATM also supports a postdoctoral program through the Advanced Studies Program at NCAR. Additional educational and outreach activities, including summer workshops and colloquia, and efforts to build digital libraries, are conducted by UCAR through partnerships with educational institutions to enhance formal and informal learning about the geosciences.

Many educational activities are undertaken as part of an individual grantee's project or as part of larger grants for small centers or university facilities. The former include involvement with K-12 students, special research and training opportunities for K-12 teachers or scientists who are involved in primarily undergraduate institutions, and public outreach activities. Examples of the latter include a two-week summer school in space weather phenomena, consequences, and modeling offered by CISM, and related summer programs are also held at the Arecibo Observatory and at the Millstone Hill Radar. Likewise, efforts associated with the CHILL Radar operated by Colorado State University give faculty and students the opportunity to explore technical and scientific topics in radar meteorology.

Because relatively few undergraduate programs offer degrees in the atmospheric sciences, the field does not benefit from the strong pipeline of students typical of other disciplines. Thus, highly talented students may be unaware of career opportunities in the atmospheric sciences or of many possible applications of training in the atmospheric sciences to other careers. While there is not a shortage of applicants for graduate studies, it is not clear that a sufficient number of top students are being attracted to the field (Vali et al., 2002). Indeed, as opportunities for science and engineering careers increase, there will be greater competition to attract talented students to the atmospheric sciences. Attracting more high-caliber students would benefit the atmospheric sciences as a whole, allowing the field to advance more quickly on many research fronts that are important to our nation and the rest of the world.

BOX 4-6
SOARS® Achievements and Successes

- Academic and workforce success:
  - 3 earned their Ph.D. in science or engineering
  - 16 currently in Ph.D. graduate programs
  - 37 earned their MS in science or engineering
  - 15 currently in MS graduate programs
  - 65 have earned undergraduate degrees in science or engineering
  - 15 are presently undergraduates
  - 24 SOARS® protégés entered workforces of NOAA, EPA, and NCAR

- Protégés honored in the field
  - 3 American Meteorological Society (AMS) Graduate Fellows
  - 3 National Science Foundation (NSF) Graduate Fellows
  - 4 NASA predoctoral fellowships

- Protégés contributing to the scientific community
  - 67 oral presentations at national or regional conferences or meetings
  - 122 posters at national or regional conferences or meetings
  - 12 refereed, protégé co-authored published papers from SOARS® research

---

The significant national investment in an excellent university infrastructure, a large national center in the atmospheric sciences, and other laboratories and institutions also warrants increased efforts to engage more extremely bright students in the atmospheric sciences. In particular, NCAR offers numerous exciting opportunities for aspiring scientists. In the past, NCAR has offered a fellowship program for graduate students; NCAR recently initiated a visitor program for graduate students, which is "designed to provide NCAR staff opportunities to bring graduate students to NCAR for 3- to 12-month collaborative visits with the endorsement of their thesis advisors and in pursuit of their thesis research" (UCAR, 2006a). Opportunities at NCAR for undergraduate and graduate students are especially valuable because so many students in the atmospheric sciences come to the field from other disciplines. A summer program near the beginning of one's graduate studies could provide an excellent orientation to the various active avenues of research in the field. The oceanography community has found this to be the case for an NSF-supported summer program at the Woods Hole Oceanographic Institution for upper-level undergraduate students.

Given the critical role of the atmospheric sciences in society's well-being, it is important to cast as wide a net as possible in attracting the next generation of atmospheric scientists. Because many colleges do not offer atmospheric science degrees, undergraduates may not be aware of the field. There are opportunities to locate talent that would not otherwise be attracted to our field, for example, students from minority-serving institutions, those with backgrounds in the liberal arts, first-generation students, and students from junior colleges. Women and those students who belong to underrepresented minority groups should be sought out in particular. The geosciences are recruiting a smaller percentage of minority students than other scientific fields. Attracting undergraduate students to summer, hands-on programs supervised by scientific mentors is a valuable pipeline for potential talent in the atmospheric sciences. Better communicating the potential career opportunities afforded by a degree in atmospheric sciences, both within the field and in other careers that require strong analytical and technical skills, may also attract students to the field. Likewise, lectures given at minority-serving institutions, liberal arts colleges, and junior colleges can help find and attract talented students who would not otherwise know about the opportunities in the atmospheric sciences.

Most U.S. atmospheric science departments are relatively small compared with the extensive subject matter that constitutes the atmospheric sciences. Also, many tools are valuable in state-of-the-art atmospheric sciences research, but they are unavailable in many atmospheric science departments. Thus graduate students in the atmospheric sciences may not have access to courses or opportunities to learn about important subject areas. The situation is particularly challenging for observational tools, as discussed above. Another example is the use of reanalysis products developed based on modern data assimilation techniques. Learning about the methods used in data reanalysis would help students understand the proper use of these reanalysis products and decrease the potential for their misuse. Other examples in the area of research tools might be graphical techniques or modern statistical methods. It would benefit the U.S. atmospheric science education effort if collaborative instructional materials were available to universities that do not have in-house capabilities to teach such material.

5

Collaborations Essential to the Atmospheric Sciences

The case studies in Chapter 2 of this report demonstrate the importance of the cross-disciplinary, interagency, international, and inter-sector aspects of those fields supported by the National Science Foundation's (NSF's) Division of Atmospheric Sciences (ATM). Collaborations are becoming increasingly important for the atmospheric sciences for several reasons. The scope of research questions has expanded, necessitating interactions with researchers from multiple other disciplines. As resources become more constrained, creative collaborations with other federal agencies and nations provide important opportunities to leverage investments in atmospheric research. Increasing societal demand for a wide range of weather, climate, and air quality forecast products and services create opportunities for collaboration with the private sector. In fact, the effective transition of research results to operational applications is a long-standing challenge for the atmospheric sciences community (NRC, 2000). In this chapter, the existing cross-disciplinary, interagency, international, and inter-sector collaborations fostered by ATM are described and opportunities for improvements are identified.

CROSS-DISCIPLINARY INTEGRATION

The National Academy of Sciences (NAS)/National Research Council (NRC) (1958) anticipated the necessity for atmospheric research to involve other disciplines, recognizing that specialists in physics, mathematics, chemistry, and engineering should join meteorologists in the new National Center for Atmospheric Research (NCAR). Indeed, around 1960, NSF agreed to include the High Altitude Observatory in the new NCAR, as a condition of Walt Roberts' becoming the first NCAR director, creating a partnership between NSF's Division of Astronomical Sciences and ATM in funding solar physics that continues today. The definition of cross-disciplinary research for atmospheric sciences has expanded substantially over the past 45 years to include biology, oceanography, economics, and societal impacts in current research. As highlighted by several of the case studies in Chapter 2 and the personal testimonials in this report, some of the highest impact and most transformative atmospheric research has taken place at disciplinary boundaries, including the discovery of and research on chaos theory, stratospheric ozone depletion, and climate change. Major efforts in climate modeling have depended upon cross-disciplinary connections.

Many challenges remain. There is a growing need for a better understanding of, for example, the linkages between chemistry, cloud microphysics, and climate; the linkages between oceans and the atmosphere; the relationship between climate and ice dynamics, including the key challenge of changes in the cryosphere; the water cycle; paleoclimate; and the health impacts of atmospheric oxidants and fine particles. In addition, cross-disciplinary aspects of the coupling between the atmosphere and the land surface, including the biosphere and the carbon cycle, remain areas of focus. Studying the climate also presents challenges to standard NSF funding mechanisms because of the long time scales of many of the phenomena. Emerging research avenues linking economics and societal impacts are of great interest, but also represent the greatest challenge insofar as their maturity and readiness must be balanced with their potential.

Aggressively pursuing cross-disciplinary research runs the risk of diverting funding from or diluting discipline-specific research. It is important to also recognize the inevitable tension between disciplinary and cross-disciplinary research. In the absence of increased funding, funding cross-disciplinary work will decrease the resources available for disciplinary research. Yet, there remain disciplinary problems which, if advances are not made, will hinder interdisciplinary research.

Effective identification of cross-disciplinary opportunities and related funding mechanisms are critical to the health of the atmospheric sciences. Yet, some research questions that fall at the interface between two or more disciplines can challenge NSF funding structures even when evaluations show these to be prime opportunities for scientific advancement. Several members of the committee, as well as many members of the broader atmospheric research community who provided input to the study, recounted anecdotal information suggesting that some cross-disciplinary research is falling between NSF's programmatic boundaries. These programmatic boundaries exist both within ATM (e.g., support for projects that straddle climate and weather research questions) and between ATM and other NSF divisions. The difficulties that exist are with finding the right program to support cross-disciplinary research projects and in harmonizing the reviews from experts in different fields. ATM leadership stressed that they collaborate with their colleagues in other divisions to support cross-disciplinary proposals and work with Principal Investigators (PIs) to identify funding opportunities. The committee believes, however, that more needs to be done to foster cross-disciplinary research. This problem cannot be solved by ATM alone, but requires also a commitment from the rest of NSF. Indeed, a recent report by the National Academy of Public Administration recommended that NSF ensure that information about cross-disciplinary research opportunities and criteria for reviewing cross-disciplinary proposals are clearly communicated to investigators (NAPA, 2004).

INTERAGENCY PROGRAMS

Several government agencies support extramural research in the atmospheric sciences—including NASA, NOAA, EPA, DOE, DoD, and FAA—in part because atmospheric science is directly relevant to the missions of these agencies. Effective coordination of ATM with other agencies is important for meeting ATM's goals for several reasons. First, many essential resources for atmospheric sciences research are created and supported by other agencies. These include space-based observational platforms, long-term monitoring efforts, and data archiving. Pooling resources supported by multiple agencies is an important component of many field programs. Second, whereas NSF's funding has remained fairly stable in recent decades, these other agencies have had more volatility. Thus, scientists supported by the other agencies turn to NSF for support when those agencies have downswings in funding, placing a larger demand on the NSF support for the atmospheric sciences. In fact, Figures 3-3 and 4-2 suggest that such a phenomenon is happening now; support for atmospheric sciences at NASA and DoD has decreased in recent years while the number of proposals received by ATM has increased. Third, because ATM is the one source for federal funding that aspires to address research needs spanning all of atmospheric science, the division has additional responsibility to consider supporting critical areas of the science not addressed by other agencies for programmatic reasons.

ATM participates in three major interagency programs that include atmospheric components (see Box 5-1): the U.S. Climate Change Science Program (CCSP), the U.S. Weather Research Program (USWRP), and the National Space Weather Program (NSWP). In addition, ATM supports the Center for Ocean, Land, and Atmosphere (COLA), a not-for-profit research institution in Calverton, Maryland, with interagency support that has some of the characteristics of the small centers discussed earlier. The

BOX 5-1
Major Interagency Programs

The **U.S. Climate Change Science Program (CCSP)** is an interagency effort to better understand how climate, climate variability, and potential human-induced changes in climate affect the environment, natural resources, infrastructure, and the economy in our nation and the world. The guiding vision for CCSP is "a nation and the global community empowered with the science-based knowledge to manage the risks and opportunities of change in the climate and related environmental systems."

The **U.S. Weather Research Program (USWRP)** has the goal of improving the delivery and use of weather information. NSF's role is to provide leadership and support for all aspects of the fundamental science components—experimental, theoretical, and numerical. The current three priority thrust areas are quantitative precipitation forecasting and estimation, hurricane landfall, and the optimal mix of observing systems.

The overarching goal of the **National Space Weather Program (NSWP)** is to achieve an active, synergistic, interagency system to provide timely, accurate, and reliable space environment observations, specifications, and forecasts. The program includes contributions from the user community, operational forecasters, researchers, modelers, and experts in instruments, communications, and data processing and analysis. It is a partnership between NSF, NASA, DoD, NOAA, DOE, the Department of the Interior, academia, and industry. NSF provides support to advance state-of-the-art instruments and data gathering techniques, to understand the physical processes, to develop predictive models, and to perform detailed analysis of data associated with past events that have caused significant impacts to space systems.

The **Center for Ocean, Land, and Atmosphere (COLA)** is devoted to understanding the predictability of Earth's current climate fluctuations on seasonal to decadal timescales using state-of-the-art, comprehensive models of the global atmosphere, world oceans, and land surface. COLA activities include (a) independently evaluating the climate variability characteristics of the nation's climate change models, (b) providing leadership on prediction of climate variability on seasonal-to-interannual time scales, (c) characterizing the impact of long-term climate change on climate variability, and (d) providing information technology infrastructure for efficient exchange of climate model and observational data. COLA is supported by NSF, NOAA, and NASA.

division contributes to these efforts by supporting scientists who are doing research on related topics and in some cases providing funds for central coordination of the programs. ATM's involvement in the CCSP, USWRP, NSWP, and COLA commits the division to ongoing support of research that addresses the goals of these programs. A possible concern has been that these targeted initiatives would constrain the community to follow certain lines of inquiry, possibly channeling emphasis away from other important research areas. However, this has not proved to be the case in the initiatives listed in Box 5-1. In fact, these initiatives have all brought new funds into ATM, thus supporting more investigators and resulting in excellent science. Many of these funds have been distributed through PI grants, and significant funds within CCSP have gone to NCAR, helping to support climate system modeling.

Interagency coordination is a long-standing challenge for federally funded research in the atmospheric sciences, as recognized in many previous reports (e.g., NRC, 1998b, 2003a), and requires the commitment of other agencies along with NSF. Yet it is essential to ensure that the critical science issues identified by the programs in Box 5-1, as well as other issues that require interagency coordination, are adequately addressed. Over the decades, interagency coordination within these programs and other interagency efforts, such as the Committee on Environment and Natural Resources Subcommittee on Air Quality Research, has exhibited mixed levels of success. The success depends in part on the leadership of each program, the willingness of the participating agencies to work toward mutual objectives, and the extent to which opportunities for coordination are clearly communicated to the research community. Typically, these interagency programs do not assert control over the budgets of individual agencies, but instead facilitate coordination by defining shared research agendas to which each agency contributes.

Interagency activities in operational meteorology and supporting research have been coordinated by the federally mandated Office of the Federal Coordinator for Meteorology (OFCM) since 1964. Fifteen federal departments and agencies currently participate in OFCM's coordination infrastructure, which includes program councils, committees, working groups, and joint action groups staffed and populated by representatives from the federal agencies. OFCM focuses on coordinating operational weather observing and forecasting requirements. In addition, it produces annual reports on federal investments in weather-related activities and research and, as needed, holds workshops and produces reports on specific issues. Like the other interagency coordination efforts, OFCM has had varied effectiveness over its tenure.

ATM is to be commended for its participation in the large interagency efforts described in Box 5-1. Furthermore, ATM program directors have been proactive about working with their colleagues from other agencies to support cross-agency research efforts (e.g., Box 5-2), in particular, field programs (see Table 4-4). The committee is concerned, however, that ATM does not appear to have a strategic approach to its interagency activities. Thus, it is not clear to the research community exactly how ATM intends to contribute to large interagency programs, and interactions between program directors from NSF and other agencies appear to have an ad hoc nature. A more strategic approach is especially important for addressing large research problems that span the research investments of multiple agencies, such as climate or air quality, and for research avenues that have significant potential applications for operational capabilities, such as weather, for which coordination with mission-oriented agencies such as the National Weather Service is critical. The inclusion of mechanisms for interagency program participation in the ATM strategic plan would both increase the transparency and decrease the ad hoc nature of NSF's approach to these interagency collaborations.

INTERNATIONAL RESEARCH ENVIRONMENT

It has long been realized that, because the atmosphere is global in extent, the meteorological discipline should span national boundaries. An International Meteorological Organization was founded in 1873 and was succeeded in 1950 by the World Meteorological Organization (WMO) organized under the umbrella of the United Nations. The WMO has fostered international cooperation on operational weather observations, for example, to ensure global coverage from satellite-based observations of the atmosphere, and has advocated free and open exchange of weather data. This cooperative international perspective has resulted in the recent establishment of international agreements for the development of a Global Earth Observing System of Systems (GEOSS; *http://earthobservation.org/*) and through international collaboration on the development of new research programs such as the World Climate Research Programme's (WCRP's) Coordinated Observation and Prediction of the Earth System (COPES; *http://copes.ipsl.jussieu.fr/index.html*), which recognizes that "there is a seamless prediction problem from weather through to climate time-scales, the necessity to address the broader climate/Earth system and the increasing ability to do this, [and] new technology for observations and computing."

Many of the major field programs over the past 50 years have involved international coordination (e.g., see Table 4-4), and several international organizations have been established to facilitate coordination of observational and other research efforts. WMO coordinated international atmospheric research programs in the past, participating in the International

Geophysical Year (1957–1958), establishing a Tropical Cyclone Project in 1971, carrying out GATE in 1974, and coordinating the GARP Global Weather and Monsoon Experiments in 1978–1979. GATE provides a good illustration of the potential complexity of international atmospheric research: it involved 40 research ships, 12 research aircraft, many moorings, and 72 countries. The WCRP was established as a successor to GARP by WMO, the International Council for Science (ICSU), and the Intergovernmental Oceanographic Commission. The WCRP has organized a succession of large projects, including the TOGA program running from 1984 to 1995; the GEWEX, which continues today; the international CLIVAR program; the study of Stratospheric Processes and their Role in Climate; the World Ocean Circulation Experiment; and the Climate and Cryosphere.

BOX 5-2
National Lightning Detection Network

Richard Orville
Department of Atmospheric Sciences
Texas A&M University, College Station

It is not widely known that Ron Taylor, NSF program director for physical meteorology, was instrumental in the start of the National Lightning Detection Network (NLDN) through the grants program in the years 1980–1983. In 1980, when I was at SUNY-Albany, Ron awarded me a grant that funded the purchase of three direction finders in the northeastern United States. The three direction finders were installed in New York in 1981 followed by two more in Pennsylvania the following year. NASA meanwhile installed a network of three direction finders in Virginia. We figured out how to connect all sensors to produce a network of eight direction finders covering the northeast in late 1982. By early 1983, the Electric Power Research Institute (EPRI) noticed our progress, reviewed our progress, and initiated annual funding to us in June 1983 at approximately $2 million. They asked us to expand our network and join with the National Severe Storms Laboratory and the Bureau of Land Management networks to cover the continental United States. The private-sector EPRI funding continued for the next six years until we completed the continental U.S. coverage in 1989 at a total investment of $12 million. And, it all started with Ron Taylor, NSF program manager, funding me as a Principal Investigator through a three-year NSF grant.

In subsequent years, the NLDN was transferred to a private company in Tucson, which was subsequently acquired by Vaisala, Inc. The network has today expanded to approximately 190 sensors and covers North America. It is a remarkable success story of cooperation between the private sector (EPRI) and the government (NSF).

A lightning sensor with the top removed showing the crossed loops that detect the azimuth to a distant lightning flash.

The International Geosphere-Biosphere Programme (IGBP) was established by ICSU to coordinate research activities on "the interactive physical, chemical, and biological processes that regulate the total Earth System, the unique environment that it provides for life, the changes that are occurring in this system, and the manner in which they are influenced by human actions" (*http://www.igbp.kva.se/*). Of particular relevance to atmospheric science, IGBP activities include the International Global Atmospheric Chemistry project, the Integrated Land Ecosystem-Atmosphere Processes Study, and the Surface Ocean-Lower Atmosphere Study. In addition, IGBP has initiated two studies to examine the Earth system as a whole: (1) Analysis, Integration and Modeling of the Earth System, which focuses on improving our understanding of the role of human perturbations to the Earth's biogeochemical cycles and their interactions with the coupled physical climate system; and (2) Past Global Changes, which is focused on understanding past climate changes.

Several activities act to coordinate modeling internationally. In part, these collaborations are directed at the assessment of climate change under the Intergovernmental Panel on Climate Change (IPCC). However, they also foster joint efforts to improve numerical models of the atmosphere and parameterizations of atmospheric processes in these models, under the aegis of international research programs such as GEWEX (e.g., the GEWEX Cloud System Study effort) and CLIVAR, and by bringing operational weather and climate modeling centers together. U.S. scientists work closely with scientists from other countries for the model execution, data analysis, and the model/data syntheses that are used to characterize the science included in assessments (e.g., IPCC, 2001) and WMO/UNEP ozone assessment reports (e.g., WMO, 2003). Models, satellite observations, and computing resources are shared across national boundaries. Atmospheric sciences has led the development of Earth system models which couple climate, oceans, land, and atmospheric chemistry, geology, and biogeochemistry. Earth system model development is now going on around the world with France, Germany, Japan, the United Kingdom, and the United States playing important roles. Many model runs are now done using ensembles of models and initial conditions to characterize uncertainties in our understanding. Model and data comparisons of models is the focus of the Working around the globe and on observational programs that are coordinated and shared internationally. Groups organized under the WCRP and WMO focus on the development and evaluation of models; for example, numerical techniques and intercomparisons of models is the focus of the Working Group on Coupled Modeling. Expanding coordination of modeling activities, forecasting, archiving of model output, and exchange of data is crucial for atmospheric sciences.

The space environment affects the entire globe, so it is not surprising that ATM research initiatives in solar-terrestrial science have a significant international dimension. The NSWP, in addition to the interagency cooperation, maintains links and collaboration to similar programs in other countries. The National Space Weather Program Implementation Plan (July 2000) specifically calls for collaboration with entities such as the International Space Environment Service and the European Space Agency. This has led to participation in workshops on space weather, such as the December 2004 European Space Weather Week, which was modeled on the highly successful annual NOAA Space Environment Center conference. The SuperDARN network of incoherent scatter radars in both the northern and southern polar regions is another example of international collaboration on the part of ATM in the area of solar-terrestrial science. Likewise,

ATM is one of 22 institutions supporting the Advanced Technology Solar Telescope under the leadership of the National Solar Observatory. ATM has also provided financial support for the International Coordination Office for the Scientific Committee On Solar-TErrestrial Physics–Climate And Weather of the Sun-Earth System (SCOSTEP-CAWSES) Program.

The U.S. atmospheric research community works within this international, intergovernmental fabric. Large field programs are discussed, planned, and approved years in advance of their going into the field. Data collected in these programs are coordinated and shared internationally. Analysis and modeling activities are also often coordinated by the United States and international steering and oversight groups of these large programs, such as CLIVAR, that work under the supervision of the WCRP. This advanced and increasing level of coordination across the nations has many benefits to all participants. However, it also creates the need for the U.S. funding agencies to make, to the extent possible, commitments of facilities, research funding, and researchers on timetables constrained by the multiple, interlocking activities of U.S. and international atmospheric scientists.

Many large international field programs are developed by international bodies, the projects of the WCRP and IGBP being especially notable in this regard, and U.S. participation is often vital to the success of these field programs. This presents a challenge to ATM because they receive proposals from U.S. investigators to participate in these field programs and, in many cases, significant budgets are involved, but at the same time the ATM budget remains relatively flat. ATM has tried to cope with this situation by knowing when such large international field programs will occur and to anticipate that some of their overall budget will be used to support the participation of U.S. investigators in these programs. There are also demands on ATM investigators to produce large numbers of IPCC climate model runs, and the NSF participation in this mainly involves NCAR staff. ATM has approached this situation in a largely ad hoc, but reasonably successful, manner so far. It is not clear that this ad hoc approach will be desired in the future when pressures on ATM funding will likely increase. A proactive and judicious mechanism, including the ability to commit with long lead time the participation of U.S. facilities and investigators, is needed for coordinated, efficient, and effective participation in international programs. Such a mechanism would help U.S. investigators and international bodies more fully understand the basis for ATM funding decisions and hence plan accordingly. In particular, this mechanism would be useful for evaluating potential ATM involvement in international field campaigns; in this case, existing international bodies (such as WCRP, the World Weather Research Program, and WMO) could help determine the merits of potential field campaigns.

The United States has been a leader in supporting atmospheric research over the past decades, but recent years have seen increasing investments, sophistication, and leadership from other nations as well. The European Union and other countries are more frequently initiating and leading major field programs. Many U.S. capabilities for observing and modeling the atmosphere and climate are matched or exceeded by Europe, the United Kingdom, and Japan. Some key examples of advances include the EU Framework programs such as ENSEMBLES, Japan's Frontier Research System for Global Change, and the European Space Agency satellite SCIAMACHY. This shift provides opportunities to leverage investments by ATM with those of other nations and also creates challenges in terms of coordinating facilities and other resources for joint studies and access to data. Indeed, the role for ATM will vary depending on the international program, ranging from taking on a leadership role or supporting international program offices to contributing to programs led by other countries.

FACILITATING COLLABORATIONS

With the increasing importance of cross-disciplinary, interagency, and international research to the advancement of the atmospheric sciences, scientists will need help to navigate interagency, intra-agency, and international boundaries and overcome the many challenges to successfully finding the support for such work. A more effective public interface and process is needed to facilitate and guide investigators seeking support of cross-disciplinary, interagency, or international research. There should be guidelines for the proposal process for these efforts.

NSF ATM's public interface, its Web site (*http://www.nsf.gov/div/index.jsp?div=ATM*), provides potential PIs information on specific, active funding opportunities. Some of these opportunities are flagged as cross-cutting and the Web pages point to a partnership of NSF program managers in and out of ATM. However, the ATM Web site does not specifically encourage or guide those who would seek to grow or obtain funding for participation in a cross-disciplinary, interagency, or international research program. It lacks any discussion of how to establish a dialog with ATM toward that end and then how links between the ATM and other divisions of NSF, other agencies or research programs in other countries should be pursued. The main Web page should provide a link to a discussion of the process, perhaps following the example set by UCAR's introduction to field project support (*http://www.ucar.edu/communications/quarterly/summer05/president.html*) that provides an explanation of the process and a generic timeline.

INTER-SECTOR COLLABORATION

As the atmospheric sciences evolve there will be increasing opportunities to exploit the skills and resources pertinent to the atmospheric sciences resident in the full range of academic, governmental, and private-sector organizations. The logical need for more intensive inter-sector collaborations arises from at least two drivers. The first is societal demand for an increasing range of weather, climate, and air quality forecast services, which can only be provided by transitioning atmospheric science research into timely data and agile models supporting operational forecasts (NRC, 2000, 2003b). The second is the increasing complexity of atmospheric sciences research, which requires ever more complex measurement systems and comprehensive computational and information management tools to meet the challenges of understanding the global atmosphere and its interactions with the biosphere, including the oceans and terrestrial surfaces as well as with solar radiation and the near-space environment (NRC, 1998b).

The role of academic researchers, supplemented by government laboratories and a few private-sector research organizations, in performing atmospheric science research in the United States is well established and their successes are widely recognized (e.g., NRC, 2003b). Government organizations have traditionally provided a range of weather and climate forecast products, now supplemented with hundreds of private-sector companies that offer diverse forecast portfolios (NRC, 2003b). The challenges of developing and operating the increasingly complex technologies required for successful atmospheric research and the need to repay society's investment in that research with a broader, more accurate, and more timely range of forecast services is opening up opportunities to engage a larger number of private-sector organizations within the atmospheric sciences. Private-sector organizations can contribute needed skills, facilities, and resources to a range of atmospheric research tasks, including instrument development; deployment and maintenance (e.g., Box 5-3); provision of commercial measurement platforms (ships, aircraft, satellites, etc.); and development, operation, and maintenance of supercomputers and other information technology tools and management systems.

In fact, as a prime consumer of supercomputer services, atmospheric and climate research centers, including NCAR, have historically had a productive relationship with corporations that develop advanced computing platforms and the software that makes them useful. As the pace of innovation in information technology quickens and computer obsolescence may be measured in months rather than years, a simple customer/vendor relationship between computationally intense atmospheric research centers and leading computer companies will seldom be appropriate.

BOX 5-3
Development of the Dropwindsonde and Radar Wind Profiler—a Public–Private Partnership

George Frederick, Strategic Development Manager
Vaisala Measurement Systems
M.S., Meteorology, University of Wisconsin at Madison

For the past 13 years I have been associated with several applications of scientific research that one way or another have benefited from NSF support. While a Senior Scientist for Radian Corporation in the 1990s we licensed an Omega dropwindsonde developed by the NSF-funded National Center for Atmospheric Research (NCAR). We adapted the sensors for commercial production, produced these instruments and marketed them to the U.S. Air Force for use as part of their airborne weather reconnaissance program. A key element of the airborne weather reconnaissance program was fixing the position and strength of tropical storms and hurricanes threatening the North American mainland and Pacific islands. The Omega dropwindsonde was the most important measurement device in this process as it measured the winds and central pressure of these storms. We also marketed these instruments to the United Kingdom and other countries as a part of their research programs. Finally, several in-house spin-off efforts were developed that resulted in adaptations of the technology for special applications. None of this would have been possible without the original NSF funding of basic NCAR research and development.

While I was still with Radian (later Radian International, and then a part of URS Corporation) we licensed through a Cooperative Research and Development Agreement the radar wind profiler technology developed at the NOAA laboratories in Boulder. NCAR was using these radars at the same time for research and had developed a number of enhancements that we later licensed. These included the improved signal processing board named PIRAQ and the enhanced signal processing algorithm, NIMA. Both of these upgrades were incorporated into the overall architecture with the assistance of NCAR scientists to provide all users the benefits of improved radar profiling technology.

Vaisala Oyj, an international company headquartered in Helsinki, Finland, acquired our instrument group from URS Corporation in 2001 and inherited the radar profiler developments mentioned above. The company also has licensed two other NCAR-developed technologies, an upgraded dropwindsonde with GPS technology and the Low Level Wind Shear Alerting System technique. Both were adapted for commercial production and subsequent sale to a wide variety of users worldwide.

Vaisala commits a significant amount of its profits to original or cooperative research and development. It also has received some matching funds from a Finnish government institute, TEKES. Taken together with the research supported by NSF, these company contributions have enabled Vaisala to maintain and grow its status as the world leader in meteorological instruments and solutions. Leveraging the NSF funding of institutions like NCAR and the research of individual scientists that contribute to the base understanding of the technology have provided the highest quality meteorological products and services available today. Vaisala's customers in turn use our products and services to help satisfy the safety and economic needs of society.

Merging organizations with different cultures and diverse goals into effective teams can be a challenging management task. While many private-sector companies share the desires to advance knowledge and serve society that motivates the best academic and government organizations, they also face the requirement to make a profit that will allow them to sustain operations and produce a return on the investments that established them. Of course, while the profit motive often affects the actions of even "nonprofit" organizations whose staffs and officers may hope to benefit financially from any intellectual property developed, it is often more compelling in a for-profit organization. Thus, while the intertwined issues of proprietary information and intellectual property can complicate relations in any research team, they are more likely to require attention in teams that include private-sector for-profit organizations.

If future atmospheric science activities are to benefit fully from inter-sector collaborations involving private-sector contributions, research team partners will need to develop agreements to define, recognize, and protect the proprietary intellectual property of each team member before the work gets started. Fortunately, many established tools—including proprietary information agreements, teaming agreements, and licensing agreements—have long been used to guide activities among private-sector organizations and can be adapted. In addition, many government organizations have developed tools, such as cooperative research and development agreements, to guide their research collaborations with other, nongovernmental organizations. However, while paper agreements can define rights and obligations, successful collaborations require a culture in which individuals understand, respect, and implement them.

The effective performance of high-level research in the atmospheric sciences and the development and delivery of the range of products that society needs enabled by that research will often require inter-sector teams of scientists and engineers. The challenge of building successful teams involving academic, government, and private-sector contributors will require significant management skills, including recognition and accommodation of cultural and motivational differences. *Fair Weather: Effective Partnerships in Weather and Climate Services* (NRC, 2003b) offers many specific recommendations for how to approach these challenges in the production of weather and climate services. Careful attention to proprietary issues, including intellectual property management, will be required. However, the potential benefits of inter-sector collaborations can greatly exceed the management challenges that will have to be met to make them effective.

6

Findings and Recommendations

In considering future directions for the atmospheric sciences, the committee reviewed the evolution of the atmospheric sciences over the past several decades, examined several examples of how National Science Foundation (NSF) support enabled major achievements in the atmospheric sciences, and analyzed the strengths and limitations of the various modes of support employed by the atmospheric division (ATM). It is clear that the division has fostered a productive research community and has been responsive to changing priorities and opportunities. On the basis of these analyses, the committee has identified the findings and recommendations discussed below. Putting several mechanisms in place to facilitate a healthy evolution of the division's activities will help ensure that this success continues. The order of the findings and recommendations presented here does not strictly reflect priorities, but rather is presented to aid the reader in following the development of the ideas presented.

PRINCIPLES FOR SUCCESSFUL SUPPORT OF THE ATMOSPHERIC SCIENCES

The committee's evaluation of ATM's evolution over the past 45 years and current activities, as discussed in Chapters 3 and 4, has revealed that the division has done a good job in meeting its mission to support the atmospheric sciences. In particular, as discussed in Chapter 2, there have been significant advances in answering fundamental scientific questions about the atmosphere, in utilizing new knowledge of the atmosphere to address societally relevant applications, and in educating a workforce to advance the science and its application. This conclusion was also the clear consensus of the many members of the broad atmospheric sciences community who have provided input to the committee's deliberations.

The committee has identified a set of 10 principles that have enabled ATM to be successful over the past 45 years. Continuing to strive to meet these principles should ensure that the division remains strong in the coming decades. A robust set of principles can be used as a framework for making funding decisions in an understandable and describable way. Such clarity is of benefit in times of expanding or declining budgets. The committee notes that all principles are not equal and that they should be applied judiciously depending upon the context.

1. **High Quality.** The division has maintained a high level of quality in the research it funds. This has been achieved through rigorous competition, strong peer review, and close working relationships between ATM program officers and members of the research community. In the case of STCs, the enforcement of a "sunset date" for the centers is generally viewed as positive, and has led to evolution that allows the centers to address cutting-edge research questions. This high level of quality is essential to the continued success of ATM.

2. **Flexibility.** ATM will be better able to meet its objectives of supporting the atmospheric science research community if it has the flexibility to apply different modes and create new modes to address evolving needs. This flexibility is essential, given the evolving roles of other federal agencies, the private sector, and the international research efforts.

3. **Responsiveness.** ATM's success over the past decades reflects in part a commitment to being responsive to the needs of the research community. Indeed, NSF's support of the atmospheric sciences is particularly important in this regard because it is the main federal agency that supports high-risk, potentially transformative research, except, of course, the National Aeronautics and Space Administration's (NASA's) satellite-based research.

4. **Balance.** Atmospheric science comprises many subdisciplines—ranging from dynamic meteorology to climate change and from atmospheric chemistry to upper atmospheric dynamics and solar physics—and is inherently interdisciplinary in that the atmosphere interacts with the oceans, land surface, and near-space environment. Furthermore, the research efforts span the spectrum from fundamental research to efforts with direct applications. A portfolio that addresses the range of these research objectives and utilizes the range of modes of support in a balanced way is essential.

5. **Interagency Partnerships.** Research in the atmospheric sciences benefits from the relevance of weather, climate, and air quality to multiple federal agencies that support some extramural research. These agencies

include NASA, the National Oceanic and Atmospheric Administration, the Department of Energy, the Environmental Protection Agency, the Federal Aviation Administration, and the Department of Defense. Building effective partnerships with other agencies that have shared priorities is critical to the long-term health of the field.

6. **Connections to International Communities.** Other nations support significant research in the atmospheric sciences, offering excellent opportunities for collaboration. ATM should maintain connections to international efforts both through engagement directly with other nations and through international programs to coordinate research (e.g., World Climate Research Programme, International Geosphere-Biosphere Programme, World Weather Research Program).

7. **Robust Research Community.** The atmospheric sciences research community includes professors and other permanent university research staff, postdoctoral fellows, graduate and undergraduate students, staff at centers (i.e., large national centers, STCs, engineering research centers), and private-sector researchers. Some stability in the support for this research community and for the training of new scientists is critical for the continuing strength of the atmospheric sciences.

8. **Community Input.** Opportunities for the broad atmospheric science community to provide input in defining strategic directions for NSF's programs help strengthen the scientific foundation of the research endeavor and build community support.

9. **Access to Necessary Resources.** The atmospheric research community needs access to appropriate observing and computational facilities. In many cases, these facilities can be shared by multiple researchers. Furthermore, resources are needed to ensure adequate time for analysis and synthesis of field campaign results.

10. **High-Quality ATM Staff.** The atmospheric sciences research community has benefited from the consistent professionalism and dedication of ATM staff over the past decades. Maintaining and renewing high-quality ATM staff with keen understanding of current scientific frontiers is essential to continued success of the field.

EMPLOYING A DIVERSITY OF MODES OF SUPPORT TO MEET ATM OBJECTIVES

The committee analyzed how each mode of support employed by ATM operates today and examined the modes that enabled several major achievements in the atmospheric sciences. Therefore, the committee concludes that each of the modes is serving an important function. In particular, the complementary roles of a large national center and grants to Principal Investigators (PIs) have been a constructive component of the atmospheric science enterprise. The diversity of available modes has facilitated several different ways to tackle the scientific questions in the atmospheric sciences. The case studies illustrate that the range of available modes has been instrumental in many of the major atmospheric sciences achievements of the last several decades. The current balance among the modes is serving the community well and the committee does not have reason to propose significant changes in balance at this time.

Another important lesson to be gleaned from our analysis of the research activities leading to the major accomplishments is that ATM has adjusted the balance from time to time as opportunities, needs, and scientific progress made necessary and possible. Indeed, it appears that many of the newer modes arose out of emerging needs of the research community. ATM may need to shift its distribution of funding modes in coming years to respond to a changing research environment. For example, domestic budget constraints at NSF and other federal agencies that support atmospheric research, increasing sophistication and investments in the international research community, and changing societal expectations of research may make it necessary to rely more on some modes of support or to introduce new modes to the ATM portfolio. Based on past experience, it is reasonable to assume that ATM will adjust the balance in the future if and when circumstances warrant.

The committee finds that the diversity of activities and modes of support is a strength of the program and of our nation's scientific infrastructure. The approach and vision outlined in NAS/NRC (1958) and the "Blue Book" ("UCAR," 1959), which together mapped out the complementary roles of a large national center and the individual investigator university grants program, has served the atmospheric science community well and is the envy of many other scientific communities. The newer modes of support, including multi-investigator awards, cooperative agreements, and centers sited at universities, complement the previously established modes. The community input received to date supports this multifaceted approach. The present balance is approximately right and reflects the current needs of the community.

RECOMMENDATION: ATM should continue to utilize the current set of modes of support for a diverse portfolio of activities (i.e., research, observations and facilities, technology development, education, outreach, and applications).

FOSTERING HIGH-RISK, POTENTIALLY TRANSFORMATIVE RESEARCH

High-risk, potentially transformative research is instrumental in making major advances in the atmospheric sciences. Thus, it is essential to continually preserve and renew opportunities for this type of research. Among federal science agencies, NSF is a leader in its commitment to support high-risk, potentially transformative research (excluding satellite instrument development). This type of research is instrumental in making major advances in the field, as illustrated by several of the cases highlighted in Chapter 2 of this report. As larger modes of support have expanded (e.g., small centers), and as peer reviewers tend to be risk averse, the opportunities for such funding are perceived as having declined. Currently, program directors have discretion to use 5 percent of their budgets for Small Grants for Exploratory Research (SGER) projects, though typically about 1 to 2 percent of each program's funds are applied this way. In addition, program directors can choose to support other high-risk work through regular grant mechanisms as they see fit. It is unknown to what extent this flexibility to support exploratory research is utilized.

The committee concludes that it is essential to create and preserve opportunities for high-risk, potentially transformative research and that the atmospheric sciences would benefit if ATM expanded its support of such projects. This would ensure that a larger portion of ATM portfolio is dedicated to supporting these research activities. It is difficult to identify specific steps to address this need, but the situation is sufficiently crucial that ATM should seek new approaches. For example, ATM might consider instituting an explicit solicitation for high-risk research, which would allow these proposals to be judged with more appropriate criteria, make it clear to the research community that the division welcomes such proposals, and ensure that program managers proactively consider supporting high-risk projects. A target of about ten such grants per year should be reasonable, although it is important to realize that opportunities for transformative research may not come every year and sometimes come in spurts. The proposal process should be kept short and the process should be as flexible as possible, encouraging excellence and innovation both in terms of the proposals and the handling by ATM management. It likely would be necessary to modify the review guidelines to explicitly reward creative, exploratory ideas and to make clear what sorts of projects would be considered high-risk and potentially transformative. Such an effort might be undertaken as a pilot program and reevaluated after several years to see if it did indeed result in breakthrough concepts frequently enough to be worth continuing.

ATM should consider other approaches to enhance opportunities for high-risk, potentially transformative research as well. For example, there may be some research questions of this type that require a bigger investment than what typically can be made by a program director or under the SGER program. One option to be more effective is to pool some of the funding for exploratory research from all ATM programs and run an internal competition to which program directors can submit promising, high-risk ideas for consideration.

RECOMMENDATION: ATM should increase the opportunities for targeted grants in support of high-risk, potentially transformative research.

ENHANCING CROSS-DISCIPLINARY, INTERAGENCY, AND INTERNATIONAL COORDINATION

The analyses of the case studies demonstrate the importance of the cross-disciplinary, interagency, and international aspects of those fields covered by ATM. Effective identification of cross-disciplinary opportunities and related funding mechanisms are critical to the health of the atmospheric sciences. Research questions in the subdisciplines of atmospheric science are interrelated. Further, many are connected to those in other scientific disciplines, such as oceanography, ecology, terrestrial science, solar physics, and social science. In some cases, the science questions extend beyond the boundaries of ATM or NSF's Geosciences directorate. ATM has fostered cross-disciplinary research, for example, by partnering with Astronomy to fund solar-terrestrial research, and by partnering with other divisions to support individual proposals or to jointly solicit proposals on a topic that falls at their interface. Yet some research questions that fall at the interface between two or more disciplines continue to challenge NSF funding structures even when evaluations show these to be prime opportunities for scientific advancement. Examples of the challenges faced in cross-disciplinary science include the need to address the water cycle, biogeochemical cycles, paleoclimate, air-sea fluxes, and health impacts of atmospheric oxidants and fine particles. Improving opportunities for cross-disciplinary research will require commitments from ATM and other NSF divisions that support related research. It is important to also recognize the inevitable tension between disciplinary and cross-disciplinary research. In the absence of increased funding, funding cross-disciplinary work will decrease the resources available for disciplinary research. Yet there remain disciplinary barriers that will hinder cross-disciplinary research if advances are not made. These considerations should be addressed by the strategic planning process discussed later in this chapter.

Despite compelling motivations for interagency coordination, ATM does not always have clear mechanisms to facilitate such interactions effectively.

Some interagency coordination takes place through formalized interagency programs (e.g., Climate Change Science Program, National Space Weather Program), interagency working groups, community-driven initiatives (e.g., Climate Variability and Change), and ad hoc interactions between program directors. A more strategic approach is needed to facilitate interagency coordination. The inclusion of mechanisms for interagency program participation in the ATM strategic plan would both increase the transparency and strengthen NSF's approach to these interagency collaborations.

The atmosphere knows no national boundaries; thus, international collaboration is critical to study of the atmosphere. The research capabilities of other nations are becoming more sophisticated and their investments in the atmospheric sciences are growing. There is a breadth of atmospheric research coordinated internationally through organizations such as the World Climate Research Programme (WCRP), International Geosphere-Biosphere Programme, World Meteorological Organization (WMO), and the Scientific Committee on Solar Terrestrial Physics. Often, these international efforts address broad cross-disciplinary research agendas. ATM has been extensively involved in international efforts, but U.S. participation has been largely on an ad hoc basis. A more strategic approach is needed to facilitate international coordination in the future especially as pressure on ATM funding increases. A proactive and judicious mechanism, including the ability to commit with long lead time the participation of U.S. facilities and investigators, is needed for coordinated, efficient, and effective participation in international programs. Such a mechanism would help U.S. investigators and international bodies more fully understand the basis for ATM funding decisions and hence plan accordingly. In particular, this mechanism would be useful for evaluating potential ATM involvement in international field campaigns; in this case, existing international bodies (such as WCRP, the World Weather Research Program, and WMO) could help determine the merits of potential field campaigns.

RECOMMENDATION: As a part of its strategic planning process, ATM should develop systematic and clearly communicated procedures for tracking international program development, identifying potential ATM contributions, committing resources where appropriate, and reevaluating participation in international activities at regular intervals.

With the increasing importance of cross-disciplinary, interagency, and international research to the advancement of the atmospheric sciences, scientists need help to navigate interagency, intra-agency, and international boundaries and overcome the many challenges to successfully finding the support for such work. NSF ATM's public interface, its Web site (*http://*

www.nsf.gov/div/index.jsp?div=ATM), provides potential PIs information on specific, active funding opportunities. Some of these opportunities are flagged as cross-cutting and the Web pages point to a partnership of NSF program managers in and out of ATM. However, the ATM Web site does not specifically encourage or guide those who would seek to grow or obtain funding for participation in cross-disciplinary, interagency, or international research program. It lacks any discussion of how to establish a dialog with ATM toward that end and then how links between the ATM and other divisions of NSF, other agencies, or research programs in other countries should be pursued. There should be guidelines for the proposal process for these efforts. The main Web page should provide a link to a discussion of the process, perhaps following the example set by UCAR's introduction to field project support (*http://www.ucar.edu/communications/quarterly/summer05/president.html*) that provides an explanation of the process and a generic timeline.

RECOMMENDATION: ATM should encourage and guide scientists seeking support to participate in cross-disciplinary, interagency, and international research by developing guidelines and procedures for initiating a dialogue about such research opportunities and then submitting formal proposals.

MEETING SUPERCOMPUTING NEEDS

The ATM-supported numerical simulation community has done a commendable job at producing high-quality research and assimilation products, given the computational constraints. How best to direct future investments in computing resources for the atmospheric sciences is a complicated issue that requires more detailed study than possible in this report. Nonetheless, the committee is convinced that good science with important societal impacts would be enabled by better, faster models, which require more powerful computers and enhanced data-storage and data-transfer capabilities. Supporting state-of-the-art computing infrastructure should be a high priority, but must be balanced by the other needs of the community so as not to jeopardize maintaining observational facilities and, especially, continued support in basic research. Meeting this demand will not likely be possible with the approaches used today and may require new organizational mechanisms, sources of funding, and partnering with other agencies, the private sector, or other nations.

RECOMMENDATION: ATM should continue to develop creative means, including interagency and international partnerships, to meet the community's demand for increased computing and data storage

capability, balancing such investments carefully against those for other research activities.

SUPPORTING FIELD PROGRAMS, DATA ARCHIVES, AND DATA ANALYSIS

ATM has well-established mechanisms for supporting short-duration field programs. However, ATM has not yet clearly articulated mechanisms for supporting field programs that require continuous, longer-term (i.e., up to multiyear) deployment and observations not available from operational monitoring networks. This type of observation protocol is generally ill suited to the existing funding opportunities, in part because they were prohibitively expensive until recently. Three factors motivate the need and appropriateness of this approach today: (1) these types of observations are especially critical to understanding the interaction between the atmosphere and Earth's surface, which is a growing area of research and concern; (2) many instruments that would be used are less expensive to operate, making it reasonable to deploy them in the field for longer durations; and (3) there are existing observational programs developed by other NSF divisions and agencies (e.g., Long Term Ecological Research, the Ocean Research Interactive Observatory Networks Ocean Observing Initiative, the proposed hydrological observatories of the Consortium of Universities for the Advancement of Hydrologic Science, Inc., and the National Ecological Observing Network), which can be leveraged with additional investments to conduct atmospheric research.

RECOMMENDATION: ATM, in coordination with other NSF divisions and federal agencies, should develop the explicit capability to support longer-term (i.e., up to multiyear) lower-atmosphere field programs to study atmospheric processes that are important on these longer time scales.

A long-standing challenge in the atmospheric sciences is providing sufficient support for scientists to analyze data obtained during field programs and from observational networks. Because analysis comes at the end of a field program and competes against the start of other new field programs, it is at times subject to reduction in support. Thus, support for field data archives, visualization tool development, and analysis is not commensurate with the investment in obtaining the measurements and the full benefit from the investment in a field program often is not realized. Maximum benefit from many NSF-supported studies also would be facilitated by easy access to data from operational observational and monitoring networks (including surface, upper air, radar, and satellite) in addition to easy access to field-program data, historical data, and numerical model data. In enhancing these capabilities, there are opportunities for NSF to work with other federal agencies that have faced similar challenges, particularly in terms of data archiving.

RECOMMENDATION: ATM should maximize the benefit of field data by ensuring that archiving, visualization, and analysis activities are well supported and continue for many years after field campaigns.

Currently, there are varied destinations for data archival, including NCAR, Web sites set up by universities, and data archives established by other government agencies (e.g., the National Climatic Data Center). For example, JOSS and the NCAR Archives have worked with the researchers on an ad hoc basis to ensure that valuable resources and datasets are stored and accessible. CEDAR maintains an archive of upper-atmosphere observations, and the Virtual Solar Observatory archives observations of the sun. However, it is becoming increasingly difficult to access older data from the standard observing network and from field programs: changing technology and analysis packages make these datasets more difficult to analyze and supporting metadata are often absent from the historical datasets. There is not always a clearly identified agency or office that has responsibility for providing archived data for researchers for both large, multi-investigator field experiments and small field experiments, and the arrangements for data management vary from experiment to experiment. Thus data archival formats, quality control, and metadata are not necessarily standardized. Furthermore, even when the data are readily available, the lack of standardized analysis software packages makes it difficult to compare analyses from different cases. At present, there is neither a formal procedure nor is their funding set aside for addressing these problems across ATM.

RECOMMENDATION: ATM should convene a committee to (a) set standards for data and metadata archival for future field programs, (b) set criteria and recommend procedures for keeping historic datasets accessible, and (c) recommend standards for software packages to enable comparison of data from different time periods.

DEVELOPING OBSERVATIONAL TOOLS

Innovative observational instruments and systems are crucial to the continued advance of atmospheric science. However, the main NSF-funding paradigm of grants to individual academic investigators is often not consistent with the wide skill sets and long time scales required for successful observational tool development and deployment. Further, it has been difficult for NCAR to sustain this capability while still maintaining and deploying its observational facilities. Given the challenges of this activity, ATM could encourage the establishment of instrument development partnerships among interested university groups, private-sector organizations, and the large center and other federally funded research and development centers (FFRDCs) in order to draw upon the full range of scientific and engineering skills and experience with field measurement requirements necessary for successful instrument and measurement systems development. FFRDC facilities as potential hosts of observational tool development activities should be assessed. The operation of other NSF instrumentation programs, including Major Research Instrumentation, Small Business Innovation Research, Small Business Technology Transfer Research, and other agency-wide, directorate, or division-level instrumentation activities, could provide possible models.

RECOMMENDATION: ATM should maintain innovative observational tool development, demonstration, and deployment as a major component of its research and development portfolio. ATM should foster new instrument development by enhancing opportunities for individual investigators to build partnerships, establish collaborative facilities, and access NCAR facilities.

Universities are becoming increasingly reluctant to invest in education programs in the observational aspects of the science for various reasons. Although there are opportunities for undergraduate and graduate students to participate in NSF-funded research projects that use or develop observational tools, it is rarer for them to be able to take courses that provide the concepts of engineering design, siting, instrument and sampling limitations, or data processing. The challenges of providing such education and training at a single university may be overcome if a more community-oriented approach is used in the development of new course materials and information technology is utilized for wide distribution of these materials. The development of good online material that can be shared nationally should be the topic of an NSF-sponsored collaboration among atmospheric science and engineering departments, the national center, the American Meteorological Society, the American Geophysical Union, the private sector, and other federal laboratories that engage in observational tool development. Such a collaboration could also select fieldwork sites that encourage hands-on engineering internships for students.

RECOMMENDATION: ATM should take concrete steps to enhance the availability of collaborative tools for university instruction in observing techniques to foster continued development of cutting-edge

instruments and to increase the general literacy among atmospheric scientists on the subject of instrumentation and observational data.

EFFECTIVELY UTILIZING CENTERS

NCAR has a rich history of collaboration with university and private-sector scientists, particularly to make progress on large scientific problems that are beyond the reach of a single university department or private-sector laboratory. Whereas there are many opportunities for collaboration between NCAR and university scientists, decisions about NCAR strategic initiatives (e.g., recent new efforts in biogeosciences and water) could benefit from broader community input. Indeed, because both NCAR and the broader atmospheric sciences community have grown in size and complexity, there are new challenges for the center in terms of maintaining a balance between inward- and outward-looking efforts. New challenges also exist in engaging a larger, more fragmented university and private-sector research community. This suggests that there may need to be additional new mechanisms to leverage the investment in a large center in a way that provides synergism with the needs of the university and private-sector research community.

Partnerships between university or private-sector scientists and existing and emerging national centers need to be strengthened. Collaborations between large national centers (both existing and emerging) and university or private-sector scientists could be enhanced by new mechanisms to stimulate joint research initiatives at a larger scale than existing ad hoc collaborations. For example, ATM could conduct a regular competition for collaborations between NCAR and the outside community, focusing on research efforts that address important atmospheric-science problems that are beyond the capability of single university departments or individual private-sector laboratories. The award should be significant, in excess of $1 million a year for five years. For initiatives that have large cross-disciplinary scope, ATM could seek mechanisms for shared funding with other NSF divisions.

RECOMMENDATION: ATM should encourage new modes of partnership between the university and private-sector research community and the large national center.

Since the late 1950s, the atmospheric research enterprise has greatly expanded to its present state in which impressive research capabilities exist in the universities, the private sector, and in federal laboratories. Even so, the fundamental rationale for a large national atmospheric sciences center outlined in the Blue Book remains valid. The national center continues to serve important objectives of the atmospheric sciences community, as articulated in its stated vision:

It is NCAR's mission to plan, organize, and conduct atmospheric and related research programs in collaboration with the universities and other institutions, to provide state-of-the-art research tools and facilities to the atmospheric sciences community, to support and enhance university atmospheric science education, and to facilitate the transfer of technology to both the public and private sectors.

The capabilities of each sector have increased tremendously since that time as have the myriad challenges and opportunities in the atmospheric sciences and allied fields. Thus, the challenge for the management of the large national center will be to prioritize and direct its activities so that it, together with the other research sectors in the atmospheric sciences, can best advance the field to the benefit of society.

RECOMMENDATION: In making choices for allocating their resources, the large national center should continue to be guided by the following mandates in consultation with and with representation from the broad U.S. atmospheric research community. It should:

1. tackle large, complex research problems, in coordination with the universities, other federal agencies, and the private sector;

2. maintain standards of scientific excellence and openness that are commensurate with its university-based mission;

3. assume a share of the leadership in the atmospheric sciences community, building on effective community collaborations;

4. provide leadership in supercomputing in support of the atmospheric sciences and the modeling of the Earth system;

5. develop community models in partnership with universities, other federal agencies, and the private sector;

6. develop advanced computational and numerical techniques and tools for use in atmospheric science;

7. enable field campaigns by coordinating their planning, managing, and logistics;

8. provide state-of-the-art archiving, access, analysis, and visualization tools for community datasets and data from NSF-sponsored field programs;

9. design, develop, and maintain state-of-the-art atmospheric instrumentation and observing platforms in partnership with universities, other federal agencies, and the private sector;

10. support education and diversity in the atmospheric sciences research community in partnership with universities, other federal agencies, and the private sector;

11. maintain vibrant postdoctoral and scientific exchange programs; and

12. foster opportunities to transfer knowledge and technology to public- and private-sector users.

These mandates are broadly in agreement with NCAR's existing mission. Over the past decades the atmospheric sciences community has carefully considered the role of the large national center and these mandates attempt to encapsulate the collective view about what its goals should be.

The few atmospheric sciences STCs (i.e., Center for the Analysis and Prediction of Storms [CAPS], Center for Clouds, Chemistry, and Climate [C4], Center for Integrated Space-Weather Modeling [CISM]) and Engineering Research Center (i.e., Collaborative Adaptive Sensing of the Atmosphere [CASA]) have contributed or are currently contributing significantly in advancing innovation and research in the atmospheric sciences. All of these small atmospheric science centers have played pivotal roles in major scientific achievements in the field that led to direct societal benefits such as improved severe storm prediction or improved space-weather forecasting. They have achieved their intended goals to: (1) support research and education of the highest quality; (2) exploit opportunities where the complexity of the research agenda requires the advantages of scope, scale, duration, equipment, and facilities that a center can provide; and (3) support innovative frontier investigations at the interfaces of disciplines and fresh approaches within disciplines. This research mode is clearly effective in advancing the science and its transition to operation.

RECOMMENDATION: ATM should actively foster opportunities for atmospheric scientists to pursue funding under the small center mode by broadly advertising the opportunity, assisting in identifying appropriate research agendas, and supporting scientists in the development of such research agendas.

RECRUITING AND TRAINING TOP STUDENTS IN THE ATMOSPHERIC SCIENCES

Recruiting and training gifted scientists is perhaps the single most important way to enable the atmospheric sciences to advance more quickly on many research fronts that are important to our nation and the rest of the world. Because relatively few undergraduate programs offer degrees in the atmospheric sciences, talented students may be unaware of career opportunities in the field. While there is no shortage of applicants for graduate studies in the atmospheric sciences, it is not clear that a sufficient number of the top students are being attracted to the field. Indeed, as opportunities for science and engineering careers increase, there will be greater competition to attract talented students to the atmospheric sciences. Given the societal importance of atmospheric science and the significant national investment in an excellent university infrastructure, a large national center, and other laboratories and institutions, the committee believes that increased efforts to attract more bright students into the field are warranted. In the past, NCAR has offered a fellowship program for graduate students. This effort could be revitalized and expanded as an ATM–universities–NCAR coopera-tive effort. Such a program could offer graduate student fellows (1) multi-year stipends similar to those for NSF graduate research fellowships and (2) a summer program, conducted jointly by NCAR and the universities near the beginning of the students' graduate studies, to acquaint students with available facilities and research opportunities. A program of this sort, sized to support about 20 new students per year at U.S. universities and advertised widely to undergraduates in related scientific majors (e.g., physics, chemistry, applied math), could be a powerful tool for recruiting top students to the atmospheric sciences.

RECOMMENDATION: ATM should establish a new university–NCAR graduate fellowship program to attract a larger share of the world's brightest students into Ph.D. programs in the atmospheric sciences.

Given the critical role of atmospheric sciences in the nation's well-being, it is important to cast as wide a net as possible in attracting the next generation of atmospheric scientists. Interestingly, the aforementioned lack of awareness of atmospheric sciences among undergraduates may provide opportunities to locate talent that would not otherwise be attracted to our field, including students from minority-serving institutions, students with backgrounds in the liberal arts, first-generation students, and students from junior colleges. Women and those students who belong to underrepresented minority groups should be sought out in particular. The geosciences are recruiting a smaller percentage of minority students than other scientific fields. Well-advertised, hands-on, summer programs supervised by scientific mentors are a valuable pipeline for potential talent in the atmospheric sci-ences. Likewise, visiting lecturer programs at minority-serving institutions, liberal arts colleges, and junior colleges attract talented students who would not otherwise know about the opportunities in the atmospheric sciences.

RECOMMENDATION: ATM should support activities that diversify the student pool by (1) continuing to support and expand research experiences for undergraduates; (2) providing opportunities for minority-serving institutions and junior colleges to partner with research universities, the large national center, and the private sector; and (3) supporting lectures given nationally by prominent atmospheric scientists.

Most U.S. atmospheric science departments are too small to fully cover the extensive subject matter that constitutes the atmospheric sciences. Also, many of the sophisticated tools used in state-of-the-art atmospheric sciences research are unavailable in many atmospheric science departments. Thus graduate students in many atmospheric sciences departments may not have access to courses or opportunities to learn about important subject areas. The situation is particularly challenging for observational tools, as discussed above, and for data assimilation techniques, graphical techniques, and modern statistical methods. It could benefit the U.S. atmospheric science education effort if collaborative instructional materials were available to universities that do not have in-house capabilities to teach such material. Some such materials may already be available, so ATM support to make them more widely accessible could also benefit atmospheric sciences education.

RECOMMENDATION: ATM should support efforts to assess the course material in U.S. atmospheric science programs, identify areas where collaborative course material could be beneficial, and fund the development of such materials for a limited number of subjects each year.

DEFINING FUTURE STRATEGIC DIRECTIONS

ATM will face continuing pressures in making decisions that impact the balance among the various modes of support. For instance, there has been, and there will continue to be, tensions about how ATM funding is allocated to the university community and the large national center. There will also be tensions about the balance between investing in major observing or computing facilities and supporting research. The committee did not find obvious problems in the balance among the various modes at present, but there are some trends that, if continued, could cause problems. One is the decrease in the fraction of single investigators versus multi-investigator proposals. Another is the increase in the fraction of funding for facilities versus research. There are good reasons for these trends and the present balance is appropriate for a healthy ATM research program today, but the implications of these trends must be periodically examined and adjustments made to ensure the long-term health of the atmospheric sciences.

A strategic plan will be essential to maintain a balanced, effective portfolio in an evolving programmatic environment. This is a time of rapid change in the demographics of graduate education, the role of the

United States in the global atmospheric science community, the role of NSF in national atmospheric science funding, and the maturation and cross-disciplinary growth of atmospheric science. It is a time of increasing challenges and opportunities in the face of constrained budgets. The atmospheric sciences community is now larger and more diverse, with an active private sector and several mission agencies along with the academic community. All three sectors are seeking improved predictions, developing new data products, and are engaged in research to some extent. Many more atmospheric observations are being taken, by more diverse platforms, including satellites, commercial aircraft, radar, and other methods. At the same time federal funding for basic research proposals in the atmospheric sciences is down. Of particular concern is decreasing funding for basic atmospheric research by other federal agencies, including NASA and the Department of Defense, forcing more and more of the community to turn to ATM for basic research funding. It is against this new context that ATM must define its role.

ATM has not published a strategic plan to guide its activities in the coming years. Given the changing programmatic environment, ATM should take a more proactive approach to strategic planning. A flexible strategic plan developed by ATM staff with ample community input will enable determination of the appropriate balance of activities and modes of support in the ATM portfolio; help plan for large or long-term investments; facilitate appropriate allocation of resources to cross-disciplinary, interagency, and international research efforts; and ensure that the United States will continue to be a leader in atmospheric research. In addition, a strategic planning effort that effectively engages the atmospheric sciences community will enhance the broad understanding of the rationale behind ATM decisions. In short, a community-based strategic planning effort could provide a *means* by which ATM can advance the preceding recommendations.

RECOMMENDATION: ATM should engage the atmospheric sciences community in the development of a strategic plan, to be revisited at regular intervals.

Strategic plans can take many different forms, ranging from describing a mission and fairly high-level goals for a program to providing more details about implementation. At a minimum the strategic plan recommended here should clearly articulate ATM's mission and goals in the context of the multidisciplinary, multiagency, and multinational environment of atmospheric research. However, the committee envisions ATM's strategic plan going beyond providing a set of goals to include actions on how to attain the goals. Although not prescribing in great detail the specifics of implementation, it should address practical implementation challenges, such as interagency relations, international relations, university, and private-sector organization relations with NCAR. Further, the plan should put flexible structures in place that will give ATM a means for making decisions about prioritization, for example, in response to pressures resulting from an evolving budgetary environment, competing international initiatives, and multiple demands for facilities. Having a strategic plan in place may call for a reorganization of ATM to direct staff and resources in a way that may better address emerging challenges.

The committee believes that the strategic plan itself will be useful to ATM, but the process of producing it may prove even more valuable, particularly if it is conducted with ample and transparent community engagement. The committee envisions the strategic planning process as providing a mechanism for the community as a whole to participate in an active conversation about the direction of the field and where best to use resources, while remaining sensitive to the societal expectations of that research. Thus, the strategic plan must be flexible and responsive, developed by the science community in collaboration with ATM management. Ideally, the process of developing the strategic plan should be straightforward and revisited at regular intervals. Furthermore, the balance of modes should evolve in the future in a manner that is consistent with strategic planning efforts.

The GEO-2000 report (GEO, 2000) represents a broad vision for the NSF Geosciences Directorate and reflects the considerable evolution of the geophysical scientific enterprise. The committee understands that GEO is revisiting its vision document and urges ATM to coordinate its efforts with those of the directorate. Indeed, the development of a strategic plan for ATM is an excellent opportunity to identify important connections with GEO and with many other parts of NSF, including the Biological Sciences Directorate, the Engineering Directorate, and the Education and Human Resources Directorate.

Many of the advances in the atmospheric sciences have been enabled by the availability of sophisticated, and expensive, facilities, including supercomputers, research aircraft, and high-power radar systems. During the past 30 years, the fraction of ATM funding devoted to facilities has grown from 23 to 33 percent of its budget. Valid arguments can, and will, be put forth for ATM purchasing bigger, and more expensive, computers and very valuable, and expensive, observing facilities in the coming years. Without significant increases in ATM's budget, purchasing these facilities will require trade-offs between investments in "tools" and funding the scientists who use them. It will be up to NSF ATM management to make these difficult decisions, and ideally this should be in the framework of their strategic planning.

Some areas of the atmospheric sciences are not presently a priority for ATM and do not receive emphasis in the form of ATM support. These areas might be (1) newly emerging, such as research efforts that exploit newly available instruments and remotely piloted vehicles; (2) subfields in which support from other agencies has in the past sustained research but has recently been greatly reduced, such as marine meteorology; or (3) areas in which ATM interest and support has waned significantly. Despite the lack of current ATM support, some of these areas are important in their own right and as a component of a well-balanced national effort in the atmospheric sciences. ATM strategic planning efforts should include a proactive approach to identifying underemphasized research areas, and responding to community needs by investing resources in currently underemphasized subdisciplines where such resources would enhance overall progress in the atmospheric sciences.

RECOMMENDATION: NSF strategic planning should consider the need to invest resources in underemphasized subdisciplines of the atmospheric sciences where the new investment would enhance progress in the atmospheric sciences.

ONGOING STRATEGIC GUIDANCE FOR ATM

The atmospheric sciences have changed greatly from the late 1950s when the NAS/NRC first assessed the status of atmospheric science research (NAS/NRC, 1958) and the "Blue Book" that guided the establishment of what is now UCAR and NCAR was written ("UCAR," 1959). There has been a significant overall expansion of federal research support for the atmospheric sciences, which in turn has led to much improved meteorological services for the U.S. public and research input into U.S. governmental and industrial decision making (see, e.g., NRC, 1998b). The infrastructure for atmospheric research has grown to be much larger and more complex than was the case 50 years ago.

The committee has concluded that the diversity of support for the atmospheric sciences is a good thing, and the balance between the various means of ATM supporting atmospheric research that now exists is reasonable. We live in a dynamic environment though. The federal funding for research ebbs and flows, institutions grow, scientific fields evolve, and scientific and technological breakthroughs can sometimes greatly accelerate the pace of change. Hence, the modes by which atmospheric research is supported need to be continually reexamined. The balance between the modes that now exists will not necessarily be correct for a future time.

Furthermore, as the national center has grown in size and scope, it is not surprising that some tension has arisen between the center and the diverse community of university and private-sector PIs. This tension was anticipated in the Blue Book ("UCAR," 1959). It is likely a natural result when a single large entity (i.e., a national center) and a collection of small entities (i.e., university and private-sector PIs) compete for the same resources in that the latter does not necessarily benefit from a unified voice and governance. In the case of the atmospheric sciences, there is the additional dimension of the desire for the national center to serve the larger atmospheric sciences community by providing observational and computational resources and opportunities for fruitful partnering.

Periodic external strategic guidance could help ATM ensure that its activities are continually evolving in a way that meets the needs of the broad atmospheric sciences community. This advice should be sought approximately every five to ten years to enable input at regular intervals. The advisory mechanism should engage the broad atmospheric sciences community, with an emphasis on obtaining balanced, objective input. Some of the issues that should be addressed include the balance and relationships among the range of scientific and societally driven research avenues, among the various modes of support employed by the division, particularly regarding potential inequities in resource distribution between large research centers or facilities and individual research scientists, and among the various subdisciplines in atmospheric research.

RECOMMENDATION: ATM should seek strategic guidance from a panel that includes representation from the fields it supports at regular intervals to ensure that its programs are well balanced and continue to meet the needs of the atmospheric sciences community.

References

Ad Hoc Committee and Technical Working Group for a Petascale Collaboratory for the Geosciences. 2005. Establishing a Petascale Collaboratory for the Geosciences: Scientific Frontiers. A Report to the Geosciences Community. UCAR/JOSS. 80 pp. Available at: *http://www.joss.ucar.edu/joss_psg/meetings/petascale/*.

AirUCI (Atmospheric Integrated Research for Understanding Chemistry at Interfaces). 2006. Available at: *http://www.chem.uci.edu/airuci/*.

AMS (American Meteorological Society). 2006. AMS Certification Programs. Available at: *http://www.ametsoc.org/amscert/index2.html#seal*.

Anderson, D. L. T., and J. P. McCreary. 1985. A note on the role of the Indian Ocean in a coupled atmosphere-ocean model of El Niño and the Southern Oscillation. J. Atmos. Sci. 42: 2439-2442.

Anderson, J. G., W. H. Brune, and M. H. Proffitt. 1989. Ozone destruction by chlorine radicals within the Antarctic vortex: The spatial and temporal evolution of ClO-O$_3$ anti-correlation based on in situ ER-2 data. J. Geophys. Res. 94: 11465-11479.

Antia, H. M., and S. Basu. 2006. Determining solar abundances using helioseismology. Astrophys. J., 644: 1292-1298.

Asplund, M., N. Grevesse, and A. Jaques Sauval. 2005. The solar chemical composition. Cosmic abundances as records of stellar evolution and nucleosynthesis, in honor of David L. Lambert. ASP Conference Series 336: 25-38.

Atlas, D. 1990. Radar in Meteorology, Boston, MA: American Meteorological Society, 806 pp. (See Chapters 21 and 28.)

Bahcall, J. N., A. M. Serenelli, and S. Basu. 2006. 10,000 standard solar models: A Monte Carlo simulation. Astropohys. J. Suppl. Ser., 165: 400.

Balsley, B. B., and K. S. Gage. 1980. The MST radar technique: Potential for middle atmospheric studies. Pure Appl. Geophys. 118: 452-493.

Barnett, T. P. 1981. Statistical prediction of North American air temperature from Pacific predictors. Mon. Wea. Rev. 9(5): 1021-1041.

Barnett, T. P. 1984. Prediction of El Niño of 1982-83. Mon. Wea. Rev. 12(7): 1403-1407.

REFERENCES

Barstow, D., and E. Geary, eds. 2002. Blueprint for Change: Report from the National Conference on the Revolution in Earth and Space Science Education. Cambridge, MA: TERC, 100 pp. Available at: *http://www.earthscienceedrevolution.org/RevEarthSciEd.pdf*.

Berger, A., J. Imbrie, J. Hays, G. Kukla, and B. Saltzman, eds. 1984. Milankovitch and Climate: Understanding the Response to Astronomical Forcing. NATO ASI Series 126 (Parts 1 and 2). Dordrecht: D. Reidel.

Bluestein, H. B. 1999a. Tornado Alley: Monster Storms of the Great Plains. New York: Oxford University Press, 180 pp.

Bluestein, H. B. 1999b. A history of storm-intercept field programs. Wea. Forecasting 14: 558-577.

Bluestein, H. B., and R. M. Wakimoto. 2003. Mobile radar observations of severe convective storms. In Radar and Atmospheric Science: A Collection of Essays in Honor of David Atlas (R. Wakimoto and R. Srivastava, eds.). Meteor. Monogr. Series 30(52), Am. Meteorol. Soc. 105-136.

Browning, K. A. 1971. Structure of the atmosphere in the vicinity of large-amplitude Kelvin-Holmholtz billows. Quart. J. Roy. Meteorol. Soc. 97: 283-299.

Burpee, R. W., J. L. Franklin, S. J. Lord, R. E. Tuleya, and S. D. Aberson. 1996. The impact of Omega dropwindsondes on operational hurricane track forecast models. Bull. Am. Meteorol. Soc. 77: 925-933.

Busalacchi, A. J., and J. J. O'Brien. 1981. Interannual variability of the equatorial Pacific. J. Geophys. Res. 86: 10901-10907.

Byers, H. R., and R. R. Braham, Jr. 1949. The Thunderstorm. U.S. Department of Commerce, 287 pp.

Cane, M., S. E. Zebiak, and S. C. Dolan. 1986. Experimental forecasts of El Niño. Nature 321: 827-832.

Cane, M. A. 1984. Modeling sea level during El Niño. J. Phys. Oceanogr. 14: 1864-1874.

Carlowicz, M. J., and R. E. Lopez. 2002. Storms from the Sun. Washington, DC: Joseph Henry Press, 234 pp.

CCSP and SGCR (Climate Change Science Program and Subcommittee on Global Change Research). 2004. Our Changing Planet: The U.S. Climate Change Science Program for Fiscal Years 2004 and 2005. Washington, DC: CCSP.

Charney, J. G., R. Fleagle, V. E. Lally, H. Riehl, and D. Q. Wark. 1966. The feasability of a global observation and analysis experiment. Bull. Am. Meteorol. Soc. 47: 200-220.

Cronin, T. M. 1999. Principles of Paleoclimatology: Perspectives in Paleobiology and Earth History Series. New York: Columbia University Press, 560 pp.

Dansgaard, W. 1964. Stable isotopes in precipitation. Tellus 16: 436-468.

Dansgaard, W. J., J. W. C. White, and S. J. Johnsen. 1989. The abrupt termination of the Younger Dryas climate event. Nature 339: 523-533.

D'Arrigo, R., R. Wilson, and G. Jacoby. 2006. On the long-term context for late twentieth century warming. J. Geophys. Res. 111: D03103 (doi:10.1029/2005JD006352).

Davies-Jones, R. P., R. J. Trapp, and H. B. Bluestein. 2001. Tornadoes and tornadic storms. In Severe Convective Storms (C. A. Doswell III, ed.). Meteor. Monogr. Series 28(50), Am. Meteorol. Soc., 167-221.

Davis, D. D., W. S. Heaps, D. Philen, M. Rodgers, T. McGee, A. Nelson, and A. J. Moriarty. 1979. Airborne laser induced fluorescence system for measuring OH and other trace gases in the part-per-quadrillion to parts-per-trillion range. Rev. Sci. Instrum 50: 1505-1516.

de Zafra, R. L., M. Jaramillo, A. Parrish, P. Solomon, B. Connor, and J. Barnett. 1987. High concentrations of chlorine monoxide at low altitudes in the Antarctic spring stratosphere: Diurnal variation. Nature 328: 408-411.

Dikpati, M., G. de Toma, P. A. Gilman, C. N. Arge, and O. R. White. 2004. Diagnostics of polar field reversal in solar cycle 23 using a flux-transport dynamo model. Astrophys. J. 601: 1136-1151.

Dikpati, M., G. de Toma, and P. Gilman. 2006. Predicting the strength of Solar cycle 24 using a flux-transport dynamo-based model. Geophys. Res. Lett. 33: L05102 (doi: 10.1029/2005GL025221).

Doswell, C. A., III, S. J. Weiss, and R. H. Johns. 1993. Tornado forecasting—A review. The Tornado: Its Structure, Dynamics, Prediction, and Hazards. Geophys. Monogr., No. 79, Amer. Geophys. Union 557-571.

Eisele, F. L., and D. J. Tanner. 1991. Ion-assisted tropospheric OH measurements. J. Geophys. Res. 96: 9295-9308.

Farman, J. C., B. G. Gardiner, and J. D. Shanklin. 1985. Large losses of total ozone in Antarctica reveal seasonal ClO$_x$/NO$_x$ interaction. Nature 315: 207-210.

Fleming, J. R. 1997. Guide to Historical Resources in the Atmospheric Sciences. NCAR, Boulder, CO. Available at: *http://www.colby.edu/sts/97guide/history.html*.

Franklin, J. L., M. L. Black, and K. Valde. 2003. GPS dropsonde wind profiles in hurricanes and their operational implications. Wea. Forecasting 18: 32-44.

Fujita, T. 1963. Analytical mesometeorology: A review. In Severe Local Storms, Meteor. Mono. 5(27), Amer. Meteor. Soc., 77-125.

Fung, I. 1986. Analysis of the seasonal and geographic patterns of atmospheric CO$_2$ distributions with a 3-D tracer model. In The Changing Carbon Cycle: A Global Analysis (J. R. Trabalka and D. E. Reichle, eds.). New York: Springer-Verlag, 592 pp.

Fung, I., C. J. Tucker, and K. C. Prentice. 1987. Application of advanced very high resolution radiometer vegetation index to study atmosphere-biosphere exchange of CO$_2$. J. Geophys. Res. 92: 2999-3015.

Gage, K. S., and B. B. Balsley. 1978. Doppler radar probing of the clear atmosphere. Bull. Amer. Meteor. Soc. 59, 1074-1093.

Geiger, R. 1957. The Climate Near the Ground. Cambridge, MA: Harvard University Press, 494 pp.

Glass, L., and M. C. Mackey. 1988. From Clocks to Chaos: The Rhythms of Life. Princeton, NJ: Princeton University Press, 272 pp.

Gleick, J. 1987. Chaos: Making a New Science. New York: Penguin Books.

GEO (National Science Foundation Geosciences Directorate) 2000: Understanding and Predicting Earth's Environment and Habitability, NSF Geosciences Beyond 2000. Science 272: 1281-1283. NSF 00-27. Arlington, VA: NSF.

Godfrey, J. S., R. A. Houze, R. H. Johnson, R. Lukas, J.-L. Redelsperger, A. Sumi, and R. Weller. 1998. Coupled Ocean-Atmosphere Response Experiment (COARE): An interim report. J. Geophys. Res. 103: 14395-14450.

Gough, D. O., J. W. Leibacher, P. H. Scherrer, and J. Toomre. 1996. Perspectives in helioseismology. Science 272: 1281-1283.

Graham, N. E., J. Michaelsen, and T. P. Barnett. 1987. An investigation of the El Niño-Southern Oscillation cycle with statistical models 1. Predictor field characteristics. J. Geophys. Res. 92: 14251-14270.

Hagan, M. 2004. Integration of atmospheric models into the ionosphere and magnetosphere. Presentation to National Academies Committee on Strategic Guidance for the National Science Foundation's Support of the Atmospheric Sciences. Boulder, CO: October 14, 2004.

Hard, T. M., R. J. O'Brien, C. Y. Chan, and A. A. Mehrabzadeh. 1984. Tropspheric Free Radical Determination by FAGE. Environ. Sci. Technol. 18: 768-777.

Harvey, J. W. 1995. Helioseismology. Physics Today 48: 32-38.

REFERENCES

Harvey J. W., F. Hill, R. P. Hubbard, J. R. Kennedy, J. W. Leibacher, J. A. Pintar, P. A. Gilman, R. W. Noyes, A. M. Title, J. Toomre, R. K. Ulrich, A. Bhatnagar, J. A. Kennewell, W. Marquette, J. Patrón, O. Saá, and E. Yasukawa. 1996. The Global Oscillation Network Group (GONG) Project. Science 272: 1284-1286.

Heiman, M., and C. D. Keeling. 1986. Meridional eddy diffusing model of the transport of atmospheric carbon dioxide. 1. The seasonal carbon cycle over the tropical Pacific Ocean. J. Geophys. Res. 91: 7765-7781.

Hoffer, T., B. Dugoni, A. Sanderson, S. Sederstrom, R. Ghadialy, and P. Rocque. 2001. Doctorate Recipients from United States Universities: Summary Report 2000. Chicago: National Opinion Research Center.

Hofmann, D. J., J. W. Harder, S. R. Rolf, and J. M. Rosen. 1987. Balloon-borne observations of the development and vertical structure of the Antarctic ozone hole in 1986. Nature 326: 59-62.

Hughes, W. J., and M. K. Hudson. 2004. Towards an integrated model of the space weather system. J. Atmos. Solar-Terr. Phys. 66(15-16): 1241-1242.

IGBP (International Geosphere-Biosphere Programme). 2003. Atmospheric Chemistry in a Changing World (G. P. Brasseur, R. G. Prinn, and A. P. Pszenny, eds.). Heidelberg, Germany: Springer-Verlag, 300 pp.

Illinois Institute of Technology, Research Institute. 1968. Technology in Retrospect and Critical Events in Science. Chicago: Illinois Institute of Technology.

Imbrie, J., and K. P. Imbrie. 1979. Ice Ages: Solving the Mystery. Cambridge, MA: Harvard University Press.

IPCC (Intergovernmental Panel on Climate Change). 2001. Climate Change 2001: The Scientific Basis. Contribution of Working Group I to the Third Assessment Report of the Intergovernmental Panel on Climate Change, J. T. Houghton, Y. Ding, D. J. Griggs, M. Noguer, P. J. van der Linden, X. Dai, K. Maskell, and C. A. Johnson, eds. Cambridge, UK: Cambridge University Press.

Keeling, C. D. 1958. The concentration and isotopic abundances of atmospheric carbon dioxide in rural areas. Geochim. Cosmochim. Acta 13(4): 322-334.

Keeling, C. D. 1960. The concentration and isotopic abundance of atmospheric carbon dioxide in the atmosphere. Tellus 12: 200-203.

Keeling, C. D. 1970. Is carbon dioxide from fossil fuel changing man's environment? Proc. Am. Philos. 114: 10-17.

Keeling, C. D. 1998. Rewards and penalties of monitoring the Earth. Annu. Rev. Energy Environ 23: 25-82.

Keeling, C. D., R. B. Bacastow, A. E. Bainbridge, C. A. Ekdahl, P. R. Guenther, and L. S. Waterman. 1976. Atmospheric carbon dioxide variations at Mauna Loa Observatory, Hawaii. Tellus 28: 538-551.

Keeling, C. D., R. B. Bacastow, A. F. Carter, S. C. Piper, T. P. Whorf, M. Heimann, W. G. Mook, and H. Roeloffzen. 1989. A three-dimensional model of atmospheric CO_2 transport based on observed winds: 1. Analysis of observational data. In Aspects of Climate Variability in the Pacific and the Western Americas, D. H. Peterson, ed. Geophys. Monogr. 55: 165-235.

Kellie, A. 2004. NSF Supported Computing Facilities. Presentation to National Academies Committee on Strategic Guidance for the National Science Foundation's Support of the Atmopsheric Sciences. Boulder, CO. October 14, 2004.

Kiehl, J. T. 2004. Goals of CCSM. Presentation to National Academies Committee on Strategic Guidance for the National Science Foundation's Support of the Atmopsheric Sciences. Boulder, CO. October 14, 2004.

Klemp, J. B. 1987. Dynamics of tornadic thunderstorms. Annu. Rev. Fluid Mech. 19: 369-402.

Kolb, C. E. 2003. Measurement challenges and strategies in atmospheric and environmental chemistry. In Challenges for the Chemical Sciences in the 21st Century—The Environment. Washington, DC: National Academy Press, 127-136.

Kovacs, T. A., and W. H. Brune. 2001. Total OH loss rate measurement. J. Atmos. Chem. 39: 105-122.

Kuo, B. 2004. The Penn State/NCAR Mesoscale Model: MM5. Presentation to National Academies Committee on Strategic Guidance for the National Science Foundation's Support of the Atmospheric Sciences. Boulder, CO. October 14, 2004.

Lanzerotti, L. J. 2003. Space weather's time has come. Space Weather 1(1): 1003 (doi:10.1029/2003SW000024).

Lau, K. M. 1981. Oscillations in a simple equatorial climate system. J. Atmos. Sci. 38: 248-261.

LeMone, M. A. 1983. The time between a field experiment and its published results. Bull. Am. Meteorol. Soc. 64: 614-615.

Lorenz, E. N. 1963. Deterministic nonperiodic flow. J. Atmos. Sci. 20: 130-141.

Lorenz, E. N. 1982. Atmospheric predictability experiments with a large numerical model. Tellus 36a: 505-513.

Lorenz, E. N. 1993. The Essence of Chaos. Seattle, WA: University of Washington Press.

Mass, C. F., and Y.-H. Kuo. 1998. Regional real-time numerical weather prediction: Current status and future potential. Bull. Am. Meteorol. Soc. 79: 253-263.

Mazuzan, G. T. 1988. Up, up, and away: The reinvigoration of meteorology in the United States: 1958-1962. Bull. Am. Meteorol. Soc. 69: 1152-1163.

McCreary, J. P. 1976. Eastern tropical ocean response to changing wind systems: With application to El Niño. J. Phys. Oceanogr. 6: 632-645.

McCreary, J. P. 1983. A model of tropical ocean-atmosphere interaction. Mon. Wea. Rev. 3 (2): 370-387.

McPhaden, M. J., and S. Hayes. 1990. Variability in the eastern equatorial Pacific Ocean during 1986-1988. J. Geophys. Res. 95: 13195-13208.

McWilliams, J., and P. Gent. 1978. A coupled air and sea model for the tropical Pacific. J. Atmos. Sci. 35: 962-989.

Mount, G. H. 1992. The measurement of tropospheric OH by long-path absorption. 1. Instrumentation. J. Geophys. Res. 97: 2427-2444.

NAPA (National Academy of Public Administration). 2004. National Science Foundation: Governance and Management for the Future. Washington, DC: NAPA.

NARSTO. 2000. An Assessment of Tropospheric Ozone Pollution—A North American Perspective. EPRI 1000040. Palo Alto, CA: EPRI.

NAS/NRC (National Academy of Sciences/National Research Council). 1958. Research and Education in Meteorology: Interim Report of the Committee on Meteorology. Washington, DC: NAS/NRC.

NCAR (National Center for Atmospheric Research). 2001. NCAR as an Integrator: A Vision for the Atmospheric Sciences and Geosciences. Boulder, CO: National Center for Atmospheric Research.

Newton, C. W. 1963. Dynamics of severe convective storms. In Severe Local Storms, Meteor. Mono. 5(27), Amer. Meteor. Soc., 33-58.

NRC (National Research Council). 1990. Status report on the Global Tropospheric Chemistry Program, prepared by the NRC Committee on Atmospheric Chemistry.

NRC. 1991. Rethinking the Ozone Problem in Urban and Regional Air Pollution. Washington, DC: National Academy Press.

NRC. 1997. Space Weather: A Research Perspective. Washington, DC: National Academy Press.

NRC. 1998a. Capacity of U.S. Climate Modeling to Support Assessments. Washington, DC: National Academy Press.

NRC. 1998b. The Atmospheric Sciences—Entering the Twenty-First Century. Washington, DC: National Academy Press.

NRC. 1999a. Adequacy of Climate Observing Systems. Washington, DC: National Academy Press.

NRC. 1999b. Global Environmental Change—Research Pathways for the Next Decade. Washington, DC: National Academy Press.

NRC. 1999c. Review of NASA's Distributed Active Archive Centers. Washington, DC: National Academy Press.

NRC. 2000. From Research to Operations in Weather Satellites and Numerical Weather Prediction: Crossing the Valley of Death. Washington, DC: National Academy Press.

NRC. 2001a. A Climate Services Vision—First Steps Toward the Future, National Research Council. Washington, DC: National Academy Press.

NRC. 2001b. Improving the Effectiveness of U.S. Climate Modeling. Washington, DC: National Academy Press.

NRC. 2002a. Abrupt Climate Change: Inevitable Surprises. Washington, DC: National Academy Press.

NRC. 2002b. Communicating Uncertainties in Weather and Climate Information: A Workshop Summary. Washington, DC: National Academy Press.

NRC. 2002c. The Sun to the Earth—and Beyond: A Decadal Research Strategy in Solar and Space Physics. Washington, DC: National Academy Press.

NRC. 2003a. Implementing Climate and Global Change Research: A Review of the Final U.S. Climate Change Science Program Strategic Plan. Washington, DC: The National Academies Press.

NRC. 2003b. Fair Weather: Effective Partnerships in Weather and Climate Services. Washington, DC: The National Academies Press.

NRC. 2004. SBIR Program Diversity and Assessment Challenges: Report of a Symposium. Washington, DC: The National Academies Press.

NRC. 2005a. Getting Up to Speed: The Future of Supercomputing. Washington, DC: The National Academies Press.

NRC. 2005b. Policy Implications of International Graduate Students and Postdoctoral Scholars in the United States. Washington, DC: The National Academies Press.

NRC. 2005c. Radiative Forcing of Climate Change: Expanding the Concept and Addressing Uncertainties. Washington, DC: The National Academies Press.

NRC. 2005d. Thinking Strategically: The Appropriate Use of Metrics for the Climate Change Science Program. Washington, DC: The National Academies Press.

NRC. 2005e. Strategic Guidance for the National Science Foundation's Support of the Atmospheric Sciences: An Interim Report. Washington, DC: The National Academies Press.

NRC. 2006a. Rising Above the Gathering Storm: Energizing and Employing America for a Brighter Economic Future. Washington, DC: The National Academies Press.

NRC. 2006b. Surface Temperature Reconstructions for the Last 2,000 Years. Washington, DC: The National Academies Press.

NSB (National Science Board). 2002. Science and Engineering Indicators—2002. NSB-02-1. Arlington, VA: National Science Foundation.

NSB. 2003. The Science and Engineering Work Force: Realizing America's Potential. NSB 03-69. Arlington, VA: National Science Foundation.

NSF (National Science Foundation). 2003. Women, Minorities and Persons with Disabilities in Science and Engineering: 2002. NSF 03-312. Arlington, VA: NSF. Available at: *http://www.nsf.gov/statistics/nsf03312/*.

NSF. 2004. Federal Funds for Research and Development, Research to Universities and Colleges by Agency and Field of Science: Fiscal Years 1973-2003. NSF 04-332 (Project Officer, Ronald L. Meeks), Arlington, VA: NSF. Available at: *http://www.nsf.gov/statistics/nsf04332/pdf/tables.pdf.*

NSF. 2005a. Policy, Procedures, and Guidelines for Science Programs That Require Field Facilities. Arlington, VA: NSF. Available at: *http://www.nsf.gov/dir_off/OFAP/info/NSFprocedures.pdf.*

NSF. 2005b. High Performance Computing System Acquisition: Towards a Petascale Computing Environment for Science and Engineering. Arlington, VA: NSF Office of Cyberinfrastructure. Available at: *http://www.nsf.gov/pubs/2005/nsf05625/nsf05625.pdf.*

NSF. 2006. Division of Science Resources Statistics, Graduate Students and Postdoctorates in Science and Engineering: Fall 2003, NSF 06-307 (Project Officer, Julia Oliver). Arlington, VA: NSF. Available at: *http://www.nsf.gov/statistics/nsf06307/pdf/nsf06307.pdf.*

NSWP (National Space Weather Program). 1995 (August). National Space Weather Program Strategic Plan, FCM-P30-1995, Washington, DC.

OFCM (Office of the Federal Coordinator for Meteorology). 2004. The Federal Plan for Meteorological Services and Supporting Research: Fiscal Year 2005. Washington, DC: OFCM.

Petit, J. R., J. Jouzel, D. Raynaud, N. I. Barkov, J.-M. Barnola, I. Basile, M. Bender, J. Chappellaz, M. Davis, G. Delayque, M. Delmotte, V. M. Kotlyakov, M. Legrand, V. Y. Lipenkov, C. Lorius, L. Pépin, C. Ritz, E. Saltzman, and M. Stievenard. 1999. Climate and atmospheric history of the past 420,000 years from the Vostok ice core, Antarctica. Nature 399: 429-436.

Prigogine, I., and I. Stengers. 1984. Order Out of Chaos. Bantam Publishers.

Prinn, R. G. 2003. The cleansing capacity of the atmosphere. Annu. Rev. Environ. Resour. 28: 29-57.

Ramanathan, V., P. J. Crutzen, J. T. Kiehl, and D. Rosenfeld. 2001a. Aerosols, climate, and the hydrological cycle. Science 294(7): 2119-2124.

Ramanathan, V., P. J. Crutzen, J. Lelieveld, A. P. Mitra, D. Althausen, J. Anderson, M. O. Andreae, W. Cantrell, G. R. Cass, C. E. Chung, A. D. Clarke, J. A. Coakley, W. D. Collins, W. C. Conant, F. Dulac, J. Heintzenberg, A. J. Heymsfield, B. Holben, S. Howell, J. Hudson, A. Jayaraman, J. T. Kiehl, T. N. Krishnamurti, D. Lubin, G. McFarquhar, T. Novakov, J. A. Ogren, I. A. Podgorny, K. Prather, K. Priestley, J. M. Prospero, P. K. Quinn, K. Rajeev, P. Rasch, S. Rupert, R. Sadourny, S. K. Satheesh, G. E. Shaw, P. Sheridan, and F. P. J. Valero. 2001b. Indian Ocean Experiment: An integrated analysis of the climate forcing and effects of the great Indo-Asian haze. J. Geophys. Res. 106(D22): 28371-28398.

Rasmussen, E. M., and T. H. Carpenter. 1982. Variations in tropical sea surface temperature and surface wind fields associated with the Southern Oscillation/El Niño. Mon. Wea. Rev. 110: 354-384.

Rotunno, R., J. B. Klemp, and R. Rotunno. 1988. A theory for strong, long-lived squall lines. J. Atmos. Sci. 45: 463-485.

Rowntree, P. 1972. The influence of tropical east Pacific Ocean temperature on the atmosphere. Quart. J. Roy. Meteor. Soc. 98: 290-321.

Ruelle, D. 1991. Chance and Chaos. Princeton, NJ: Princeton University Press, 214 pp.

Schopf, P. S., and M. J. Suarez. 1987. Vacillations in a coupled ocean-atmosphere model. J. Atmos. Sci. 45: 549-566.

Serafin, R., R. B. Heikes, D. Sargeant, W. Smith, E. Takle, D. Thomson, and R. Wakimoto. 1991. Study on observational systems: A review of meteorological and oceanographic education in observational techniques and the relationship to national facilities and needs. Bull. Am. Meterol. Soc. 72: 815-826.

REFERENCES

Shukla, J., and J. M. Wallace. 1983. Numerical simulation of the atmospheric response to equatorial Pacific sea surface temperature anomalies. J. Atmos. Sci. 40: 1613-1630.

Simpson, S. 2004. A sun-to-mud education in two weeks. Space Weather 2: S07002 (doi:10.1029/2004SW000092).

Siscoe, G. 2006. A culture of improving forecasts: Lessons from meteorology. Space Wea. 4: S01003 (doi:10.1029/2005SW000178).

Solomon, S., G. H. Mount, R. W. Sanders, and A. L. Schmeltekopf. 1987. Visible spectroscopy at McMurdo station, Antarctica, 2, Observations of OClO. J. Geophys. Res. 92: 8329-8338.

Stevens, P. S., J. H. Mather, and W. H. Brune. 1994. Measurement of OH and HO_2 by laser-induced-fluorescence at low pressure. J. Geophys. Res. 99: 3543-3557.

Takle, E. S. 2000. University instruction in observational techniques: Survey responses. Bull. Am. Meteorol. Soc. 81: 1319-1325.

"UCAR" (University Committee on Atmospheric Research). 1959. Preliminary Plans for a National Institute for Atmospheric Research (the "Blue Book"). Prepared for the National Science Foundation under Grant G 5807. Available at: http://www.ucar.edu/news/images/PDFs/BlueBook.PDF. Accessed July 29, 2005.

UCAR (University Corporation for Atmospheric Research). 1986. Global Tropospheric Chemistry—Plans for a U.S. Research Effort. UCAR Office of Interdisciplinary Earth Sciences. OIES Report 3. Boulder, CO.

UCAR. 2005. Field Program Support at UCAR Report of the Field Program support committee. Available at: http://www.ucar.edu/fps/fps.pdf.

UCAR. 2006a. Advanced Study Program. Available at: http://www.asp.ucar.edu/graduate/graduate_visitor.jsp.

UCAR. 2006b. Windows to the Universe. Available at: http://www.windows.ucar.edu.

UCAR Quarterly. 2003. NSF center boosts space weather modeling. Available at: http://www.ucar.edu/communications/quarterly/winter02/modeling.html.

Vali, G., and R. Anthes. 2003. UCAR 2003 Graduate Student Enrollment Survey. Available at: http://www-das.uwyo.edu/~vali/grad_stud/surv03.html. Accessed August 30, 2005.

Vali, G., R. Anthes, D. Thomson, D. Houghton, J. Fellows, and S. Friberg. 2002. Wanted: More Ph.D.s graduate enrollment in the atmospheric sciences. Bull. Am. Meteorol. Soc. 83: 63-71.

Vincent, R. A., and I. M. Reid. 1983. HF Doppler measurements of mesospheric gravity wave momentum fluxes. J. Atmos. Sci. 40: 1321-1333.

Wang, C. C., L. I. Davis, Jr., P. M. Selzer, and R. Munoz. 1981. Improved airborne measurements of OH in the atmosphere using the technique of laser-induced fluorescence. J. Geophys. Res. 86: 1181-1186.

WebCASPAR (Web-based Computer-Aided Science Policy Analysis and Research). 2006. Integrated Science and Engineering Research Data System. Available at: http://webcaspar.nsf.gov/.

Weisman, M. L., and C. A. Davis. 1998. Mechanisms for the generation of mesoscale vortices within quasi-linear convective systems. J. Atmos. Sci. 55: 2603-2622.

Wilhelmson, R. B., and L. J. Wicker. 2001. Numerical modeling of severe local storms. Severe Convective Storms. Meteor. Monogr. 28(50) (C. Doswell III, ed.), Amer. Meteor. Soc. 123-166.

WMO (World Meteorological Organization). 1989. Scientific Assessment of Stratospheric Ozone: 1989. Global Ozone Research and Monitoring Project—Report No. 20, Geneva, Switzerland: WMO.

WMO. 2003. Scientific Assessment of Ozone Depletion: 2002. Global Ozone Research and Monitoring Project—Report No. 47, Geneva, Switzerland: WMO.

Woodman, R. F., and A. Guillen. 1974. Radar observations of winds and turbulence in the stratosphere and mesosphere. J. Atmos. Sci. 31: 493-505.

Wyrtki, K. 1975. El Niño—The dynamic response of the equatorial Pacific Ocean to atmospheric forcing. J. Phys. Oceanogr. 5: 572-584.

Zebiak, S. E. 1982. A simple atmospheric model of relevance to El Niño. J. Atmos. Sci. 39: 2017-2027.

Zebiak, S. E., and M. A. Cane. 1987. A model El Niño Southern Oscillation. Mon. Wea. Rev. 115: 2262-2278.

A

Statement of Task

At the request of the Division of Atmospheric Sciences (ATM), this committee will perform a study that will provide guidance to ATM on its strategy for achieving its goals in the atmospheric sciences (e.g., cutting-edge research, education and workforce development, service to society, computational and observational objectives, data management). In doing so, the committee will seek to engage the broad atmospheric science community to the fullest extent possible. The committee will provide guidance on the most effective approaches for different goals and on determining the appropriate balance among approaches. In essence, the committee is asked to consider how ATM can best accomplish its mission of supporting the atmospheric sciences into the future. Specifically, this study will consider the following questions:

1. What are the most effective activities (e.g., research, facilities, technology development, education and workforce programs) and modes of support (e.g., individual principal investigators, university-based research centers, large centers) for achieving the National Science Foundation's (NSF's) range of goals in the atmospheric sciences?

2. Is the balance among the types of activities appropriate and should it be adjusted? Is the balance among modes of support for the atmospheric sciences effective and should it be adjusted?

3. Are there any gaps in the activities supported by the Division and are there new mechanisms that should be considered in planning and facilitating these activities?

4. Are interdisciplinary, foundation-wide, interagency, and international activities effectively implemented and are there new mechanisms that should be considered?

5. How can NSF ensure and encourage the broadest participation and involvement of atmospheric researchers at a variety of institutions?

The study will not make budgetary recommendations. The committee will deliver its results in two parts: (1) a short interim report in fall 2005 that provides a preliminary sense of the committee's overarching conclusions; and (2) a final report by fall 2006 that further considers community input and provides the committee's full analysis and recommendations.

B

Biographical Sketches of
Committee Members and Staff

COMMITTEE MEMBERS

Dr. John A. Armstrong, *Chair*, received his Ph.D. in the field of nuclear magnetic resonance from Harvard University in 1961. Dr. Armstrong spent most of his career at IBM, until he retired as vice president of science and technology. He is the author or co-author of some 60 papers on nuclear resonance, nonlinear optics, the photon statistics of lasers, picosecond pulse measurements, the multiphoton spectroscopy of atoms, the management of research in industry, and issues of science and technology policy. As a result of his contributions in nonlinear optics, quantum physics, and technical leadership in advanced very-large-scale integration technology, Dr. Armstrong was elected a member of the National Academy of Engineering (NAE) in 1987. In addition, he received the George E. Pake Prize of the American Physical Society (APS) in 1989. Dr. Armstrong was a member of the presidentially appointed National Advisory Committee on Semiconductors. He was also a member of the National Science Board from 1996 to 2002 and served on its Special Commission on the Future of the National Science Foundation (NSF). Dr. Armstrong has served on numerous National Research Council (NRC) bodies, including the Commission on Physical Sciences, Mathematics, and Applications, where he was liaison to the Computer Science and Technology Board; he chaired the Committee on Partnerships in Weather and Climate Services and the Committee on Future Needs in Deep Submergence Science. Dr. Armstrong serves as chair of the Industrial Advisory Board for the NSF Engineering Research Center

"Collaborative Adaptive Sensing of the Atmosphere" located at University of Massachusetts at Amherst.

Dr. Susan K. Avery is vice chancellor and dean at the University of Colorado, Boulder. Dr. Avery received her Ph.D. in atmospheric science from the University of Illinois. She served as assistant professor in the Department of Electrical Engineering, University of Illinois, Urbana; professor in the Department of Electrical and Computer Engineering, University of Colorado; and associate dean of research and graduate education, College of Engineering, University of Colorado, Boulder, and director of the Cooperative Institute for Research in Environmental Sciences. Dr. Avery has broad interests in upper-atmosphere dynamics, Doppler-radar techniques for observing the atmosphere, and the application of weather and climate information for decision support. From 2002 to 2003, she assisted the U.S. Climate Change Science Program in drafting its strategic plan; she was particularly instrumental in shaping the chapter on decision support, a new emphasis for the program. Dr. Avery is the current president of the American Meteorological Society (AMS), a member of the American Geophysical Union (AGU), and a Fellow of the Institute of Electrical and Electronics Engineers. She is a past officer of the University Corporation of Atmospheric Research (UCAR). Dr. Avery receives research support from NSF and the National Oceanic and Atmospheric Administration (NOAA).

Dr. Howard B. Bluestein is a George Lynn Cross Research Professor of Meteorology at the University of Oklahoma, where he has served since 1976. He received his Ph.D. in meteorology from the Massachusetts Institute of Technology. His research interests are the observation and physical understanding of weather phenomena on convective, mesoscale, and synoptic scales. Dr. Bluestein is a fellow of the AMS and of the Cooperative Institute for Mesoscale Meteorological Studies. He is chair of the NSF Observing Facilities Advisory Panel, the past chair of the AMS Committee on Severe Local Storms and UCAR's Scientific Program Evaluation Committee, a past member of the AMS Board of Meteorological and Oceanographic Education in Universities, and a former member of the NRC Board on Atmospheric Sciences and Climate (BASC). He is also the author of a textbook on synoptic-dynamic meteorology and of *Tornado Alley*, a book for the scientific layperson on severe thunderstorms and tornadoes. Dr. Bluestein receives research support from NSF.

Dr. Elbert W. (Joe) Friday is the WeatherNews Chair Emeritus of Applied Meteorology and Director Emeritus of the Sasaki Institute at the University of Oklahoma. Most recently, he was director of the NRC BASC from 1998 to 2002, and senior scholar from 2002 to 2003. In the previous year, he served as the assistant administrator for research for NOAA. From 1988 to 1997, he was director of the National Weather Service, serving during its extensive modernization. During this same period, he served as the U.S. Permanent Representative to the World Meteorological Organization. Dr. Friday completed a 20-year career in the U.S. Air Force, retiring in 1981 as a colonel. He is a fellow and past president of the AMS and a member of the American Association for the Advancement of Sciences, the National Weather Association, and the research society Sigma Xi. He has been awarded the Presidential Rank Award of Meritorious Executive, the Distinguished Graduate Award from the University of Oklahoma, where he received a Ph.D. in Meteorology in 1969, and the 1993 Federal Executive of the Year Award from the Federal Executive Institute Alumni Association. He received the 1997 Cleveland Abbe Award for Outstanding Service from AMS.

Dr. Marvin A. Geller is a professor of atmospheric sciences at the State University of New York at Stony Brook. His research deals with atmospheric dynamics, middle and upper atmosphere, climate variability, and aeronomy. Dr. Geller has served on many national and international advisory committees on atmospheric science, the upper atmosphere, and the near-space environment, and is currently president of the Scientific Committee on Solar-Terrestrial Physics (SCOSTEP); the NSF's Division of Atmospheric Sciences pays dues to SCOSTEP through BASC. His past NRC service includes a 2003 Review of the National Aeronautics and Space Administration (NASA) Earth Sciences Enterprise Strategic Plan; membership on BASC, the Committee on Metrics for Global Change Research, and the Committee on Solar and Space Physics; and chair of the Committee on Solar-Terrestrial Research. He is a fellow of AMS, a fellow of AGU, and past president of AGU's Atmospheric Sciences Section. Dr. Geller receives research funding from NSF and NASA.

Dr. Elisabeth A. Holland obtained her Ph.D. from Colorado State University in 1988, followed by a postdoctoral fellowship at Stanford University. She has worked in the Atmospheric Chemistry Division at the National Center for Atmospheric Research (NCAR) since 1989, focusing on linkages between atmospheric chemistry and terrestrial ecosystems. She has combined modeling and measurements to examine interactions between the terrestrial carbon and nitrogen cycles, ranging from initial endeavors in microbiology to her current focus on global and regional biogeochemistry. Dr. Holland directed the NATO Advanced Study Institute on Soils and Global Change, was an associate editor for the *Journal of Geophysical Research*, a fellow with both the Natural Resources Ecology Laboratory (Colorado State University) and the Institute for Arctic and Alpine Ecology

(University of Colorado), and serves on a number of steering committees including the International Committee on Atmospheric Chemistry and Global Pollution and NCAR's Significant Opportunities in Atmospheric and Related Sciences program, which provides research opportunities to minority students. Dr. Holland is a member of the graduate faculty at Colorado State University and the University of Colorado, and has also worked with students from Stanford University, the University of California at Berkeley, State University of New York, Stony Brook, and the University of New Hampshire. From 1999 to 2001, she was C3 Professor and Atmospheric Chemistry Group Leader for the Max Planck Institute of Biogeochemistry in Jena, Germany.

Dr. Charles E. Kolb received his Ph.D. from Princeton University in physical chemistry. Dr. Kolb is president and chief executive officer of Aerodyne Research, Inc. in Billerica, Massachusetts. Aerodyne is a private company that receives research support from many government agencies, including NSF. Dr. Kolb's principle research interests have included atmospheric and environmental chemistry, combustion chemistry, materials chemistry, and the chemical physics of rocket and aircraft exhaust plumes. He has served on several NASA panels dealing with environmental issues, as well as on several previous NRC committees and boards dealing with atmospheric and environmental chemistry. These include the NRC's BASC, the Committee to Review NARSTO's Scientific Assessment of Airborne Particulate Matter, and the Committee on Atmospheric Chemistry. He is a fellow of the APS, AGU, the American Association for the Advancement of Science, and the Optical Society of America.

Dr. Margaret A. LeMone is a senior scientist at NCAR. She has two primary scientific interests: (1) the structure and dynamics of the atmosphere's planetary boundary layer and its interaction with the underlying surface and clouds overhead; and (2) the interaction of mesoscale convection with the boundary layer and surface underneath, and with the surrounding atmosphere. Dr. LeMone is also the chief scientist for Global Learning through Observations for the Benefit of the Environment (GLOBE), a worldwide hands-on, primary- and secondary-school-based science and education outreach program. GLOBE is operated by UCAR and Colorado State University under a cooperative agreement with NASA. GLOBE also receives in-kind support from the State Department; NSF funds PIs to help oversee and provide quality control for GLOBE measurements and to use GLOBE data in their research. Dr. LeMone's salary is supported in part by NCAR and in part by the GLOBE program. Dr. LeMone is a fellow of the American Association for the Advancement of Science and AMS. She is also a member of NAE and a former member of BASC. She has served on the NRC's Panel on Improving the Effectiveness of U.S. Climate Modeling, the Special Fields and Interdisciplinary Engineering Peer Committee of the NAE, and the Committee on Weather Research for Surface Transportation. Dr. LeMone received her Ph.D. in atmospheric sciences from the University of Washington.

Dr. Ramón E. López received his Ph.D. in space physics in 1986 from Rice University. He is a professor of physics and space sciences at the Florida Institute of Technology. Prior to this appointment, he was the C. Sharp Cook Distinguished Professor in the Department of Physics at the University of Texas at El Paso. Dr. López is a fellow of the APS and was awarded the 2002 APS Nicholson Medal for Humanitarian Service. In 2003, he was elected vice chair of the APS Forum on Education and to serve as chair in 2005. Dr. López leads a research group that is working in both space physics and science education. His current research focuses on making detailed quantitative comparisons between the results of global three-dimensional magnetohydrodynamic simulations and observations during actual events, as well as student interpretation of visualizations. Dr. López receives research support from NASA and NSF. He is the author or co-author of 86 scientific publications and 18 nonscientific publications, including the popular science book *Storms from the Sun*. From 1994 to 1999, he was director of Education and Outreach Programs of APS. Dr. López is active in science education reform nationally. He has served as an education consultant for a number of school districts around the country, for state education agencies in California, Maryland, North Carolina, and Texas, and for federal agencies such as NASA and the NSF; and he was a member the NRC's Committee on Undergraduate Science Education.

Dr. Susan Solomon is widely recognized as one of the leaders in the field of atmospheric science. Since receiving her Ph.D. degree in chemistry from the University of California at Berkeley in 1981, she has been employed by NOAA as a research scientist. She made some of the first measurements in the Antarctic that showed that chlorofluorocarbons were responsible for the stratospheric ozone hole, and she pioneered the theoretical understanding of the surface chemistry that causes it. In March 2000, she received the National Medal of Science, the United States' highest scientific honor, for "key insights in explaining the cause of the Antarctic ozone hole." Her current research focuses on chemistry-climate coupling, and she serves as co-chair of Working Group I of the Intergovernmental Panel on Climate Change, which seeks to provide scientific information to the United Nations Framework Convention on Climate Change. Dr. Solomon was elected to the National Academy of Sciences in 1992.

Dr. John M. Wallace is a professor of atmospheric sciences at the University of Washington. His research has improved our understanding of global climate and its interannual and decadal variations, through the use of observational data. He has been instrumental in identifying and understanding a number of atmospheric phenomena such as the spatial patterns in month-to-month and year-to-year climate variability, including the one through which the El Niño phenomenon in the tropical Pacific influences climate over North America. Dr. Wallace receives research support from NSF and NOAA. Dr. Wallace is a member of the National Academy of Sciences and has chaired several NRC panels including the Panel on Reconciling Temperature Observations, the Panel on Dynamic Extended Range Forecasting, and the Advisory Panel for the Tropical Ocean/Global Atmosphere (TOGA). He has also served on committees addressing Abrupt Climate Change: Implications for Science and Public Policy, and the Science of Climate Change.

Dr. Robert A. Weller received his Ph.D. in 1978 from Scripps Institution of Oceanography. He is the director of the Cooperative Institute for Climate and Ocean Research at Woods Hole Oceanographic Institution (WHOI), and has worked at WHOI since 1979. His research focuses on atmospheric forcing (wind stress and buoyancy flux), surface waves on the upper ocean, prediction of upper ocean variability, and the ocean's role in climate. He has served as the Secretary of the Navy Chair in Oceanography. He has been on multiple mooring deployment cruises and has practical experience with ocean observation instruments. Dr. Weller receives research support from NOAA, the Naval Research Laboratory, and NSF. He is currently a co-chair of the U.S. Climate Variability and Change (CLIVAR) Scientific Steering Group and a member of the international CLIVAR Scientific Steering Group. CLIVAR receives funding from NSF's ATM and Ocean Science Division. Dr. Weller has served on several NRC committees over the years, including the recent Committee to Review the U.S. Climate Change Science Program Strategic Plan and the Committee on Implementation of a Seafloor Observatory Network for Oceanographic Research; he was also a member of BASC. He is currently serving on the NRC Committee on Utilization of Environmental Satellite Data: A Vision for 2010 and Beyond.

Dr. Stephen E. Zebiak is director-general, as well as director of Modeling and Prediction Research, at the International Research Institute (IRI) for climate prediction, hosted at Columbia University. IRI is supported by NOAA, the U.S. Agency for International Development, the Department of Energy, NSF, and international sources. Dr. Zebiak has worked in the area of ocean-atmosphere interaction and climate variability since completing his Ph.D. in 1984. He was an author of the first dynamical model used to predict El Niño successfully. He has served as chair of the International CLIVAR Working Group on Seasonal to Interannual Prediction, co-chair of the U.S. CLIVAR Seasonal to Interannual Modeling and Prediction Panel, and member of numerous advisory committees for U.S. and international science programs. He has served as a member of AMS's Committee on Climate Variations, and as an associate editor for the *Journal of Climate*. Dr. Zebiak's expertise with intermediate-scale climate models and the interpretation of ocean and atmospheric modeling outputs on decadal and interannual scales will provide an important input to this study. Dr. Zebiak was a member of the NRC Advisory Panel for the Tropical Ocean and Global Atmosphere program and the Committee on Improving the Effectiveness of U.S. Climate Modeling.

NRC STAFF

Dr. Amanda C. Staudt is a senior program officer with BASC. She received an A.B. in environmental engineering and sciences and a Ph.D. in atmospheric sciences from Harvard University. Her doctorate research involved developing a global three-dimensional chemical transport model to investigate how long-range transport of continental pollutants affects the chemical composition of the remote tropical Pacific troposphere. Since joining the National Academies in 2001, Dr. Staudt has staffed the National Academies review of the U.S. Climate Change Science Program Strategic Plan and the long-standing Climate Research Committee. Dr. Staudt has also worked on studies addressing radiative forcing of climate, surface temperature reconstructions, air quality management in the United States, research priorities for airborne particulate matter, the NARSTO Assessment of the Atmospheric Science on Particulate Matter, weather research for surface transportation, and weather forecasting for aviation traffic flow management.

Dr. Claudia Mengelt is a program officer for BASC. After completing her B.S. in Aquatic Biology at the University of California, Santa Barbara, she received her M.S. in Biological Oceanography from the College of Oceanic and Atmospheric Sciences at Oregon State. Her Master's research focused on how chemical and physical parameters in the surface ocean effect Antarctic phytoplankton species composition and consequently impact biogeochemical cycles. She obtained her Ph.D. in the Marine Sciences from the University of California, Santa Barbara, where she conducted research on the photophysiology of harmful algal species. She joined the full-time staff of BASC in the fall of 2005 following a fellowship with the NRC Polar Research Board in the winter of 2005. At the National Academies, she has worked on studies addressing the design of Arctic observing systems and evaluating lessons learned from global change assessments.

Dr. Curtis Marshall is a program officer for BASC. He received B.S. (1995) and M.S. (1998) degrees in meteorology from the University of Oklahoma, and a Ph.D. (2004) in Atmospheric Science from Colorado State University. His doctoral research examined the impact of anthropogenic land-use change on the mesoscale climate of the Florida peninsula. Prior to joining the staff of BASC in 2006, he was employed as a research scientist at NOAA, where he focused on the development of coupled atmosphere–land surface models.

Ms. Elizabeth A. Galinis is a research associate for BASC. After completing her B.S. in marine science from the University of South Carolina in 2001, she received her M.S. in environmental science and policy from Johns Hopkins University in 2006. Since her start at the National Academies in March 2002, Ms. Galinis has worked on studies involving next-generation weather radar (NEXRAD), weather modification, climate sensitivity, climate change, radiative forcings, the Global Energy and Water Cycle Experiment Americas Prediction Project, U.S. future needs for polar icebreakers, and the effects of climate change on federal lands.

C

Individuals Who Provided Input to the Committee

Over the past two years, the committee has met six times to gather information and conduct deliberations. At several of these meetings, members of the atmospheric sciences community were invited to share their perspectives on study questions, both in sessions devoted to specific issues and in an "open mike" session when any comments were welcome. In addition, the committee made available a Web site through which members of the community could contribute comments (*http://dels.nas.edu/basc/strat.shtml*), met with the heads and chairs of the University Corporation for Atmospheric Research (UCAR) universities, and held town hall sessions at the December 2004 fall meeting of the American Geophysical Union (AGU) and at the January 2005 annual meeting of the American Meteorological Society (AMS). The committee's interim report was made available online and was mailed to the heads and chairs of the UCAR universities with instructions on how to submit comments and the announcements of upcoming town halls. Two town halls were held at the AGU meeting, fall 2005, and the AMS meeting, winter 2006, to solicit input from the community in response to its charge and its interim report. The committee has also received input through the above-mentioned Web site in response to the interim report. This input has been quite helpful in shaping the committee's thinking. We acknowledge in particular the following individuals who made substantive comments in one or more of these venues:

M. Joan Alexander, NorthWest Research Associates
Caspar Amman, National Center for Atmospheric Research (NCAR)
Richard Anthes, UCAR

Dave Atlas, NASA Goddard Space Flight Center
Kile B. Baker, National Science Foundation (NSF)
Robert C. Beardsley, Woods Hole Oceanographic Institution
Richard Behnke, NSF
Rosina Bierbaum, University of Michigan
Aaron Brasket, Energy Velocity, Boulder, Colorado
Rob Brue, University of California, Berkeley
Richard Carbone, NCAR
Frederick H. Carr, University of Oklahoma
Carol A. Clayson, Florida State University
Ron Cohen, University of California, Berkeley
Walter Dabberdt, Vaisala, Inc.
Ben de Foy, Massachusetts Institute of Technology (MIT)
Terry Deschler, University of Wyoming
Kelvin Droegemeier, University of Oklahoma
Kerry Emanuel, MIT
Jay Fein, NSF
Jack Fellows, UCAR
Carl Friedman, New Mexico Institute of Mining & Technology
John Gaynor, National Oceanic and Atmospheric Administration (NOAA)
Peter Gilman, NCAR
Maura Hagan, NCAR
Chuck Hakkarinen, Electric Power Research Institute, *Retired*
Kevin Hamilton, University of Hawaii
Dennis Hartmann, University of Washington
Ernest Hildner, Space Environment Center, NOAA
David Hofmann, Climate Monitoring and Diagnostics Laboratory, NOAA
Clifford Jacobs, NSF
Roberta Johnson, UCAR
Peter Kallay, University of California, Davis
Al Kellie, NCAR
Jeff Kiehl, NCAR
Timothy Killeen, NCAR
Joe Klemp, NCAR
Michael Knoekler, NCAR
Paul Krehbiel, New Mexico Institute of Mining & Technology
Bill Kuo, NCAR
David Legler, CLIVAR Project Office
Peter Leavitt, Weather Information, Inc.
Doug Lilly, University of Oklahoma
Roland List, University of Toronto
Jennifer Logan, Harvard University

Brian Mapes, University of Miami
Denise Mauzerall, Woodrow Wilson School, Princeton University
R. C. Mercure, Jr., CEO of a technology-based company
Christopher Mooers, University of Miami
Jarvis Moyers, NSF
Sandy MacDonald, Forecast Systems Laboratory, NOAA
Chris McCormick, Broad Reach Engineering, Boulder, Colorado
Danny McKenna, NCAR
Natalie Mahowald, NCAR
William Neff, Environmental Technology Laboratory, NOAA
John W. Nielsen-Gammon, Texas A&M
Raj Pandya, UCAR
Dave Parsons, NCAR
Annick Pouquet, NCAR
Lynn Preston, NSF
V. Ramanathan, Scripps Institution of Oceanography and University of
    California, San Diego
Roy Rasmussen, NCAR
Gene Rasmusson, University of Maryland, College Park
Alan Robock, Rutgers University
Steve Rutledge, Colorado State University
Cindy Schmidt, UCAR
Bob Serafin, NCAR
Steven Sherwood, Yale University
John Snow, University of Oklahoma
Tim Spangler, NCAR
Paul Sperry, Cooperative Institute for Research in Environmental Sciences
Pamela Stephens, NSF
Gene Takle, Iowa State University
Bruce Umminger, NSF
Gabor Vali, University of Wyoming
Susan VanGundy, National Science Digital Library
Tom Vonder Haar, Colorado State University
Roger Wakimoto, NCAR
Robert M. White, Washington Advisors Group
Don Wuebbles, University of Illinois
Xubin Zeng, University of Arizona

D

Acronyms

AAP   Atmospheric Analysis Prediction
ABC   Atmospheric Brown Clouds
ACD   Atmospheric Chemistry Division
ACE-1   First Aerosol Characterization Experiment
ACSYS   Arctic Climate System Study
AFGL   U.S. Air Force Geophysics Laboratory
AFWA   Air Force Weather Agency
AGU   American Geophysical Union
AIMES   Analysis, Integration, and Modeling of the Earth System
AirUCI   Environmental Molecular Sciences Institute (EMSI) at UC
    Irvine
ALPEX   ALPine EXperiment
AMISR   Advanced Modular Incoherent Scatter Radar
AMMA   African Monsoon Multiscale Analysis
AMS   American Meteorological Society
APS   American Physical Society
ASAP   Automated Shipboard Aerological Program
ASR   Annual Scientific Reports
AST   Antarctic Submillimeter Telescope
ATD   NCAR Atmospheric Technology Division
ATM   NSF Division of Atmospheric Sciences
ATS-1   First Geosynchronous Satellite
AVAPS   Automatic Vertical Atmospheric Profiling System

BAMEX	Bow echo and Mesoscale convective EXperiment
BGC WG	BioGeoChemistry Working Group
BiSON	Birmingham Solar Oscillation Network
BOREAS	BOreal Ecosystem-Atmosphere Study
C4	Center for Clouds, Chemistry, and Climate
CAPS	Center for the Analysis and Prediction of Storms
CAREER	NSF faculty early career development program
CASA	Collaborative Adaptive Sensing of the Atmosphere
CAWSES	Climate And Weather of the Sun-Earth System
CCSM	Community Climate System Model
CCSP	Climate Change Science Program
OCD	Office of Climate Dynamics
CEDAR	Coupling, Energetic, and Dynamics of Atmospheric Regions
CEPEX	CEntral Pacific EXperiment
CFC	Chlorofluorocarbon
CHILL	Colorado State University Chill National Radar Facility
CIMS	Chemical Ionization Mass Spectrometry
CINDE	Convective INitiation and Downburst Experiment
CISM	Center for Integrated Space-weather Modeling
CLIMAP	Climate: Mapping, Analysis, and Prediction
CLIVAR	CLImate VARiability and change
COARE	Coupled Ocean Atmosphere Response Experiment
COHMAP	Co-Operative Holocene MApping Project
COLA	Center for Ocean, Land, and Atmosphere
COMET	COoperative Meteorological Education and Training
CON3	the average of forecasts from 3 models—HRD's barotropic VICBAR model, NCEP's global spectral model, and the GFDL hurricane model
COPES	Coordinated Observation and Prediction of the Earth System
COPS	Cooperative Oklahoma Profiler Studies
COSMIC	Constellation Observing System for Meteorology, Ionosphere, and Climate
CPT	Climate Process and modeling Teams
CRADA	Cooperative Research and Development Agreements
CSM	Climate System Model
CSEM	Center for Space Environmental Modeling
CUAHSI	Consortium of Universities for the Advancement of Hydrologic Science, Inc.
CW	Continuous Wave

DLR   Deutsche Luft-und Raumfahrt (German Aerospace Research Establishment)

DoD   Department of Defense

DOE   Department of Energy

DOT   Department of Transportation

DOW   Doppler-on-Wheels

DSP   Dynamical Seasonal Prediction

ECHO   Experiment for Coordinated Helioseismic Observations

ECI   Environmental Change Institute

ECMWF   European Centre for Medium-Range Weather Forecasts

ELDORA   ELectra DOppler RAdar

EMSI   Environmental Molecular Sciences Institute

ENSEMBLES   Ensemble-based Predictions of Climate Changes and their Impacts

ENSO   El Niño/Southern Oscillation

EOL   NCAR Earth Observing Laboratory

EOS   Earth Observing System

EPA   Environmental Protection Agency

EPIC   Eastern Pacific Investigation of Climate

EPRI   Electric Power Research Institute

ERBE   Earth Radiation Budget Experiment

ERC   Engineering Research Center

ERDA   Energy Research and Development Agency

ERICA   Experiment on Rapidly-Intensifying Cyclones over the Atlantic

ESA   European Space Agency

ESH   Earth System History

ESSL   NCAR Earth and Sun Systems Laboratory

EU   European Union

FAA   Federal Aviation Administration

FASTEX   Fronts and Atlantic Storm Track EXperiment

FFRDC   Federally Funded Research and Development Center

FGGE   First GARP Global Experiment

FSL   Forecast System Laboratory

FY   Fiscal Year

GAIM   Global Analysis, Integration, and Modeling

GALE   Genesis of Atlantic Lows Experiment

GALLEX   Gallium Radiochemical Solar Neutrino Detector at Gran Sasso

GAOS   Global Atmospheric Observing System

GARP	Global Atmospheric Research Program
GATE	GARP Atlantic Tropical Experiment
GEM	Geospace Environment Modeling
GEO	NSF Geosciences Directorate
GEOSS	Global Earth Observing System of Systems
GEWEX	Global Energy and Water cycle EXperiment
GISS	Goddard Institute for Space Studies
GLOBE	Global Learning through Observations for the Benefit of the Environment
GONG	Global Oscillation Network Group
GPS	Global Positioning System
GRF	Graduate Research Fellowships
GWE	Global Weather Experiment
HAO	High Altitude Observatory
HDH	High-Degree Helioseimometer
HIAPER	High-performance Instrumented Airborne Platform for Environmental Research
ICSU	International Council for Science
IGAC	International Global Atmospheric Chemistry Project
IGBP	International Geosphere-Biosphere Programme
IHOP_2002	International H$_2$O Project
iLEAPS	Integrated Land Ecosystem-Atmosphere Processes Study
INDOEX	INDian Ocean EXperiment
IPCC	Intergovernmental Panel on Climate Change
IRI	International Research Institute
ISLSCP	International Satellite Land Surface Climatology Project
ISS	Integrated Sounding System
ITR	International Technology Research
JOSS	Joint Office for Science Support
KDI	Knowledge and Distributed Intelligence
KISS	Kanton Island Sounding System
L2D2	Lightweight Loran Digital Dropsonde
LCF	DOE's Leadership-Class Computing Facility
LEAD	Linked Environments for Atmospheric Discovery
LIDAR	LIght Detection And Ranging
LIF	Laser Induced Fluorescence
LLWAS	Low Level Wind Shear Alerting System
LOD2	Lightweight Omega Digital Dropsondes

LORAN — LOng RAnge Navigation
LOWL — Low- and Intermediate-Degree Experiment

MALT — Mesosphere and Lower Thermosphere
MAP — Mesoscale Alpine Program
MCV — Mesoscale Convective Vortices
MIT — Massachusetts Institute of Technology
MIRAGE-MEX — Megacity Impacts on Regional and Global Environments: Mexico City Pollution Outflow Field Campaign
MLSO — Mauna Loa Solar Observatory
MM — Mesoscale Model
MM4 — Mesoscale Model-version 4
MM5 — Mesoscale Model-version 5
MMM — Mesoscale and Microscale Meteorology
MONEX — MONsoon EXperiment
MRI — Major Research Instrumentation
MST — Mesosphere-Stratosphere-Troposphere Radar
MURI — Department of Defense Multidisciplinary Research Program

NAME — North American Monsoon Experiment
NAPA — National Academy of Public Administration
NARSTO — North American Research Strategy for Tropospheric Ozone
NAS — National Academy of Sciences
NASA — National Aeronautics and Space Administration
NCAR — National Center for Atmospheric Research
NCEP — National Centers for Environmental Prediction
NEXRAD — NEXt generation RADar
NHRE — National Hail Research Experiment
NIMA — NCAR Improved Moments Algorithm
NIO — Northern Indian Ocean
NJIT — New Jersey Institute of Technology
NLDN — National Lightning Detection Network
NOAA — National Oceanic and Atmospheric Administration
NRC — National Research Council
NSB — National Science Board
NSF — National Science Foundation
NSO — National Solar Observatory
NSSL — National Severe Storms Laboratory
NSWP — National Space and Weather Program
NTP — National Toxicology Program
NWP — Numerical Weather Prediction

OCE	Ocean Sciences
ODWs	Omega DropWindsondes
OFCM	Office of the Federal Coordinator for Meteorology
OGD	Ogden, Utah
OPP	Office of Polar Programs
ORION OOI	Ocean Research Interactive Observatory Networks Ocean Observing Initiative
OU	University of Oklahoma
PAGES	PAst Global changES
PECASE	Presidential Early Career Awards for Scientists and Engineers
PI	Principal Investigator
PIRAQ	PC-Integrated Radar AcQuisition system
PRESTORM	Oklahoma–Kansas Preliminary Regional Experiment for STORM–Central
PROVOST	PRediction Of climate Variations On Seasonal and inter-annual Timescales
RAMS	Regional Atmospheric Modeling System
RASS	Radio Accoustic Sounding System
REU	Research Experience for Undergraduates
RFP	Request for Proposal
RICO	Rain in Cumulus over the Oceans
RUI	Research at Undergraduate Institutions
S-Pol	S-Band Dual Polarization Radar
SAGE	Stratospheric Aerosol and Gas Experiment
SBIR	Small Business Innovation Research
SCIAMACHY	SCanning Imaging Absorption spectroMeter for Atmospheric CHartographY
SCOSTEP	Scientific Committee On Solar TErrestrial Physics
SEC	NOAA Space Environment Center
SGCR	Subcommittee on Global Change Research
SGER	Small Grants for Exploratory Research
SHEBA	Surface HEat Budget of the Arctic Ocean
SHINE	Solar and Heliospheric Interaction
SOARS®	Scientific Opportunities in Atmospheric and Related Sciences
SOHO	SOlar and Heliospheric Observatory
SOLAS	Surface Ocean-Lower Atmosphere Study
SPARC	Stratospheric Processes And their Role in Climate
SRI	Stanford Research Institute

STC   Science and Technology Center
STEPS   Severe Thunderstorm Electrification and Precipitation Study
STORM-central   Oklahoma-Kansas PRE-STORM Program
STR   Solar-Terrestrial Research program
STRATO2C   German high-altitude research aircraft
STTR   Small Business Technology Transfer
SUNY   State University of New York
SuperDARN   Super Dual Auroral Radar Network

T-Rex   Terrain-induced Rotor EXperiment
TAO   Tropical Atmosphere Ocean Observing System
TAOS   Tethered Atmospheric Observing Systems
TEKES   The Finish National Technology Agency
THERMEX   THERMal wave EXperiment
THORPEX   THe Observing system Research and Predictability EXperiment

TOGA   Tropical Ocean and Global Atmosphere
TOPSE   Tropospheric Ozone Production about the Spring Equinox

TOTO   TOtable Tornado Observatory

UCAR   University Corporation for Atmospheric Research
"UCAR"   University Committee for Atmospheric Research
UHF   Ultrahigh Frequency
UN   United Nations
UNEP   United Nations Environmental Programme
UOP   UCAR Office of Programs
URS   United Research Services
USDA   United States Department of Agriculture
USGCRP   U.S. Global Climate Research Program
USGS   United States Geological Survey
USSR   Union of Soviet Socialist Republics
USWRP   U.S. Weather Research Program

VHF   Very High Frequency
VORTEX   Verification of the ORigin of Tornadoes EXperiment

WACCM   Whole Atmosphere
WCRP   World Climate Research Programme
WGCM   Working Group on Coupled Modeling
WMO   World Meteorological Organization
WOCE   World Ocean Circulation Experiment

APPENDIX D

WP²	Western Pacific Warm Pool
WRF	Weather Research and Forecast
WWRP	World Weather Research Program